Geometry
NEXT GENERATION
Workbook

2024-25
Donny Brusca

ISBN 978-1-952401-44-2

© 2024 CourseWorkBooks, Inc. All rights reserved.

www.CourseWorkBooks.com

Table of Contents

	Unit	Topic	Page
Chapter 1.	**Basic Geometry**		**5**
	1.1	Lines, Angles and Shapes	5
	1.2	Pythagorean Theorem	11
	1.3	Perimeter and Circumference	14
	1.4	Area	18
Chapter 2.	**Coordinate Geometry**		**26**
	2.1	Forms of Linear Equations	26
	2.2	Parallel and Perpendicular Lines	28
	2.3	Distance Formula	31
	2.4	Midpoint Formula	34
	2.5	Perpendicular Bisectors	36
	2.6	Directed Line Segments	39
Chapter 3.	**Polygons in the Coordinate Plane**		**41**
	3.1	Triangles in the Coordinate Plane	41
	3.2	Quadrilaterals in the Coordinate Plane	46
	3.3	Perimeter and Area using Coordinates	52
Chapter 4.	**Rigid Motions**		**57**
	4.1	Translations	57
	4.2	Line Reflections	61
	4.3	Rotations	66
	4.4	Point Reflections	70
	4.5	Carry a Polygon onto Itself	72
Chapter 5.	**Dilations**		**74**
	5.1	Dilations of Line Segments	74
	5.2	Dilations of Polygons	77
	5.3	Dilations of Lines	81
Chapter 6.	**Transformation Proofs**		**82**
	6.1	Properties of Transformations	82
	6.2	Sequences of Transformations	85
	6.3	Transformations and Congruence	91
	6.4	Transformations and Similarity	93
Chapter 7.	**Circles in the Coordinate Plane**		**95**
	7.1	Equation of a Circle	95
	7.2	Graph Circles	99
Chapter 8.	**Foundations of Euclidean Geometry**		**101**
	8.1	Postulates, Theorems and Proofs	101
	8.2	Parallel Lines and Transversals	105

Chapter 9.   Triangles and Congruence 109
        9.1   Angles of Triangles   109
        9.2   Triangle Inequality Theorem   114
        9.3   Segments in Triangles   117
        9.4   Isosceles and Equilateral Triangles   121
        9.5   Triangle Congruence Methods   126
        9.6   Prove Triangles Congruent   130
        9.7   Overlapping Triangles   135

Chapter 10.   Triangles and Similarity 139
        10.1   Properties of Similar Triangles   139
        10.2   Triangle Similarity Methods   143
        10.3   Prove Triangles Similar   145
        10.4   Triangle Angle Bisector Theorem   148
        10.5   Side Splitter Theorem   150
        10.6   Triangle Midsegment Theorem   154

Chapter 11.   Points of Concurrency 157
        11.1   Incenter and Circumcenter   157
        11.2   Orthocenter and Centroid   160

Chapter 12.   Right Triangles 164
        12.1   Congruent Right Triangles   164
        12.2   Equidistance Theorems   167
        12.3   Geometric Mean Theorems   170

Chapter 13.   Trigonometry 174
        13.1   Trigonometric Ratios   174
        13.2   Use Trigonometry to Find a Side   177
        13.3   Use Trigonometry to Find an Angle   182
        13.4   Special Triangles   186
        13.5   Cofunctions   187
        13.6   SAS Sine Formula for Area of a Triangle   189

Chapter 14.   Quadrilaterals 193
        14.1   Angles of Polygons   193
        14.2   Properties of Quadrilaterals   195
        14.3   Trapezoids   198
        14.4   Use Quadrilateral Properties in Proofs   200
        14.5   Prove Types of Quadrilaterals   204

Chapter 15.   Circles 207
        15.1   Circumference and Rotation   207
        15.2   Arcs and Chords   209
        15.3   Tangents   213
        15.4   Secants   216
        15.5   Circle Proofs   220
        15.6   Arc Lengths and Sectors   224

Chapter 16. Solids **229**
      16.1 Volume of a Sphere   229
      16.2 Volume of a Prism or Cylinder   230
      16.3 Volume of a Pyramid or Cone   234
      16.4 Density   236
      16.5 Lateral Area and Surface Area   238
      16.6 Rotations of Two-Dimensional Objects   241
      16.7 Cross Sections   244

Chapter 17. Constructions **246**
      17.1 Copy Segments, Angles, and Triangles   246
      17.2 Construct an Equilateral Triangle   249
      17.3 Construct an Angle Bisector   251
      17.4 Construct a Perpendicular Bisector   255
      17.5 Construct Lines Through a Point   258
      17.6 Construct Inscribed Regular Polygons   261
      17.7 Construct Points of Concurrency   263
      17.8 Construct Circles of Triangles   265

Chapter 1.   Basic Geometry

1.1   Lines, Angles and Shapes

MODEL PROBLEM 1: LINES AND ANGLES

$\overline{AB} \perp \overline{CD}$, m∠$BEC = (3x+3)°$, $AE = x+1$ cm, and $EB = 2x+12$ cm. Find AB.

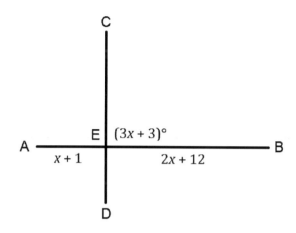

Solution:
- (A) $3x + 3 = 90$
- (B) $3x = 87$
  $x = 29$
- (C) $AE = 29 + 1 = 30$ cm
  $EB = 2(29) + 12 = 70$ cm
- (D) $AB = AE + EB = 100$ cm

Explanation of steps:
- (A) Perpendicular lines form right angles *[m∠BEC = 90°]*.
- (B) Solve for x.
- (C) Substitute for x in the expressions.
- (D) The length of a segment is equal to the sum of its parts *[AB = AE + EB]*.

PRACTICE PROBLEMS

1. Which of the following has two endpoints?

   (1) angle
   (2) line
   (3) line segment
   (4) ray

2. Which name refers specifically to ∠1, marked by an arc in this diagram?

   (1) ∠A
   (2) ∠CAD
   (3) ∠BAD
   (4) ∠ACD

   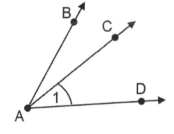

3. A diagram of $\overline{WZ}$ is shown below. Which represents the same line segment as $\overline{WZ}$?

(1) $\overline{WX}$
(2) $\overline{XY}$
(3) $\overline{WXY}$
(4) $\overline{ZYXW}$

4. An angle bisector always divides an angle into two

(1) congruent angles
(2) right angles
(3) complementary angles
(4) supplementary angles

5. An angle that measures more than 90° and less than 180° is

(1) acute
(2) right
(3) straight
(4) obtuse

6. If ∠1 and ∠2 are complementary angles and m∠1 = 20°, then

(1) m∠2 = 20°
(2) m∠2 = 70°
(3) m∠2 = 90°
(4) m∠2 = 160°

7. If two congruent angles are supplementary, each angle measures

(1) 45°
(2) 90°
(3) 50°
(4) 180°

8. If ∠1 and ∠2 are vertical angles, then which of the following is *not* true?

(1) The angles share the same vertex.
(2) The angles are adjacent.
(3) The angles are congruent.
(4) The angles are formed by intersecting lines.

Basic Geometry   1.1 Lines, Angles and Shapes

9. In the diagram, $\overleftrightarrow{DG} \perp \overleftrightarrow{EH}$ and the two lines intersect at C. Ray $\overrightarrow{CF}$ is drawn. Which of the following statements is false?

    (1) ∠ECF and ∠FCG are complementary
    (2) ∠ECF and ∠FCH are a linear pair
    (3) ∠ECF and ∠DCH are a pair of vertical angles
    (4) ∠ECD and ∠DCH are right angles

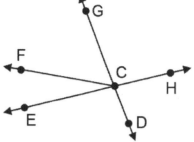

10. In the diagram, $\overline{DE}$ bisects $\overline{AC}$ at B. Which of the following statements must be true?

    (1) B is the midpoint of $\overline{DE}$
    (2) $\overline{DB} \cong \overline{BE}$
    (3) AC = DE
    (4) $BC = \frac{1}{2} AC$

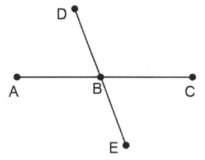

11. In the diagram, B is the midpoint of $\overline{AC}$, $AB = 2x - 40$ and $BC = x + 10$. Find AC.

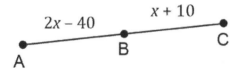

12. In the diagram, ∠ABC is a straight angle, m∠ABD = $(6x + 5)°$ and m∠DBC = $x°$. Find x.

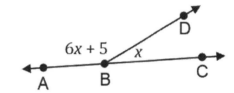

13. In the diagram, ∠BAD is a right angle, m∠BAC = $(x+15)°$, and m∠DAC = $(x-5)°$.
Find m∠BAC and m∠DAC.

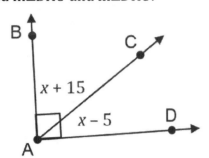

14. In the diagram, $\overrightarrow{AC}$ is the bisector of ∠BAD. If m∠CAD = $(4x+1)°$ and m∠BAD = 75°, find x.

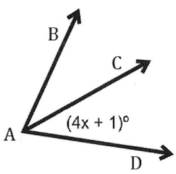

MODEL PROBLEM 2: POLYGONS

In the diagram below, name one triangle and two different quadrilaterals.

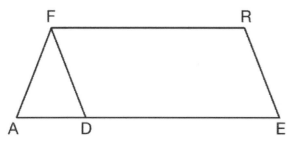

Solution:
Triangle *AFD* and quadrilaterals *FRED* and *FREA*

Explanation of steps:
Polygons are named by their vertices in either clockwise or counterclockwise order.
[There is one triangle shown in the diagram. Listing the vertices in clockwise order, the triangle may be named AFD, FDA, or DAF. In counterclockwise order of its vertices, the triangle may be named ADF, DFA, or FAD. Any of these six names may be used to specify the triangle.
There are two different quadrilaterals in the diagram. The smaller one may be given any of the following names: FRED, REDF, EDFR, or DFRE in clockwise order or FDER, DERF, ERFD, or RFDE in counterclockwise order. These are the only eight names that may be used for this polygon.
The larger quadrilateral, which is a composite figure made up of the triangle and smaller quadrilateral joined together, may be given any of the following names: FREA, REAF, EAFR, or AFRE in clockwise order or FAER, AERF, ERFA, or RFAE in counterclockwise order.]

PRACTICE PROBLEMS

15. If the triangle below is to be named by its vertices, what are the six possible names that it may be given?

16. Which of the following names may *not* be used to describe the quadrilateral shown below?

    (1) MRPG   (3) PRGM
    (2) RPGM   (4) MGPR

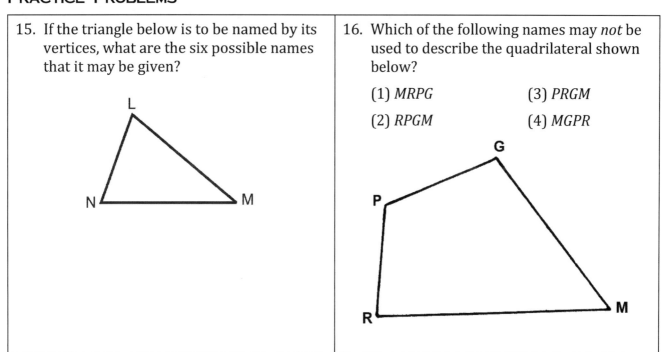

17. In quadrilateral *ABCD* below, draw and name the diagonals.

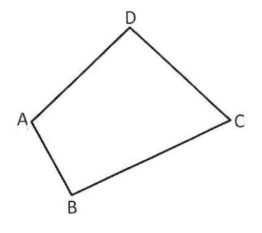

18. In the diagram below, *E* is a point on $\overline{AD}$.

    a) Find and name 3 different triangles.

    b) Find and name 3 different quadrilaterals.

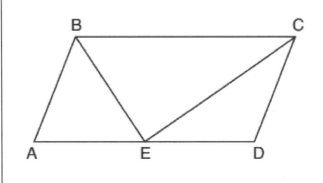

19. In the diagram below, *U* is a point on $\overline{TS}$, *V* is a point on $\overline{QR}$, and $\overline{UV}$ and $\overline{TR}$ intersect at *W*.

    a) find and name 4 different triangles

    b) find and name 5 different quadrilaterals

    c) find and name 2 different pentagons

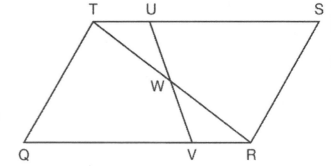

Basic Geometry 1.2 Pythagorean Theorem

1.2 Pythagorean Theorem

Model Problem

A wall is supported by a brace 10 feet long, as shown in the diagram to the right. If one end of the brace is placed 6 feet from the base of the wall, how many feet up the wall does the brace reach?

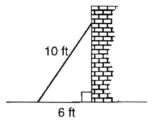

Solution:
(A) $a^2 + b^2 = c^2$
(B) $6^2 + b^2 = 10^2$
(C) $36 + b^2 = 100$
$b^2 = 64$
$b = \sqrt{64} = 8$ ft.

Explanation of steps:
(A) Given two sides of a right triangle, use the Pythagorean Theorem to find the third side.
(B) Substitute given legs as a and b (in either order), and substitute the hypotenuse, if given, as c.
[leg a = 6 and hypotenuse c = 10]
(C) Solve for the remaining variable, and simplify the radical if possible. When taking the square root of both sides, ignore the negative square root since the length of a side must be positive.

Practice Problems

1. If the lengths of the legs of a right triangle are 5 and 7, what is the length of the hypotenuse?	2. The hypotenuse of a right triangle is 26 centimeters and one leg is 24 centimeters. Find the number of centimeters in the second leg.

3. A ladder is placed against a wall as shown in the diagram below. What is the distance, *x*, from the foot of the ladder to the base of the wall, to the *nearest tenth of a foot*?

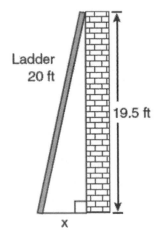

4. An 18-foot ladder leans against a wall. The base of the ladder is 9 feet from the wall on level ground. How many feet up the wall, to the *nearest tenth of a foot*, is the top of the ladder?

5. A 10-foot ladder is placed against the side of a building as shown in Figure 1 below. The bottom of the ladder is 8 feet from the base of the building. In order to increase the reach of the ladder against the building, it is moved 4 feet closer to the base of the building as shown in Figure 2. To the *nearest tenth of a foot*, how much further up the building does the ladder now reach?

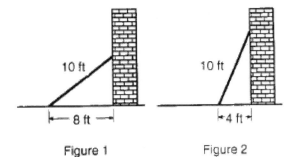

6. A kite with two support bars is shown in the diagram below. The sides of the kite measure 7 inches each on top and *x* inches each on the bottoms, meeting at right angles as shown. The vertical bar is (*x* + 1) inches. What is the measure, in inches, of the vertical bar?

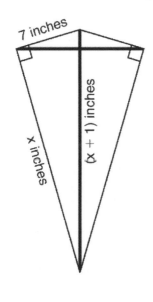

Basic Geometry   1.3 Perimeter and Circumference

1.3 Perimeter and Circumference

MODEL PROBLEM

The "key" in a regulation size basketball court is made up of a rectangle and semicircle. The length of the rectangle measures 19 feet from the baseline to the free throw line, and the width measures 12 feet, as shown below. Find the total distance around the key to the *nearest foot*.

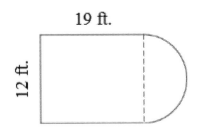

Solution:

(A) $P = 2l + w + \frac{1}{2}\pi d$

(B) $= 2(19) + (12) + \frac{1}{2}\pi(12)$

(C) $= 50 + 6\pi \approx 69$ feet

Explanation of steps:
(A) Write an equation that adds the parts that make up the distance around the <u>outside</u> of the composite figure. *[The outside of this figure uses only 3 sides of a rectangle – two lengths and a width – and half the circumference of a circle.]*
(B) Substitute the given values for the variables *[19, 12, and 12, for l, w and d]*.
(C) Simplify *[and use a calculator to find the sum rounded to the nearest foot]*.

PRACTICE PROBLEMS

1. The second side of a triangle is two more than the first side, and the third side is three less than the first side. Write an expression, in simplest form in terms of *x*, for the perimeter of the triangle.	2. The plot of land illustrated in the accompanying diagram has a perimeter of 34 yards. Find the length, in yards, of *each* side of the figure. 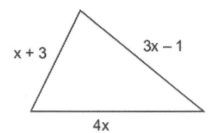

14

3. The figure below consists of a semicircle with a radius of 4 inches and a rectangle with a width of 7 inches. Find the perimeter to the *nearest tenth of an inch*.

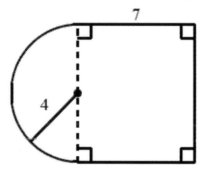

4. The figure below consists of a semicircle with radius 2 cm and a rectangle with a length of 4 cm. Find the perimeter to the *nearest tenth of a cm*.

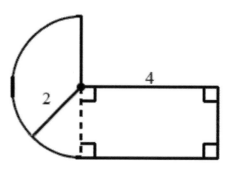

5. The figure below consists of a rectangle and a semicircle. Find the perimeter of the figure to the *nearest tenth of a foot*.

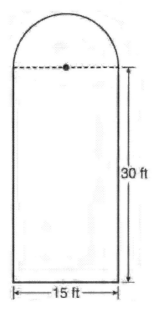

6. The figure below consists of a square with a side of 4 cm, a semicircle on the top, and an equilateral triangle on the bottom. Find the perimeter of the figure to the *nearest tenth of a centimeter*.

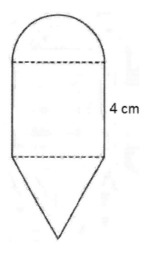

7. What is the perimeter of the figure shown below, which consists of an isosceles trapezoid and a semicircle?

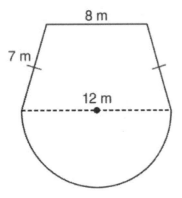

8. What is the perimeter of the figure shown below, which consists of an isosceles trapezoid and a semicircle?

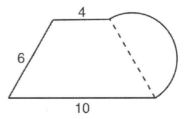

9. The polygon below consists of two squares and an equilateral triangle. The length of $\overline{AB}$ is 3.5 inches. Find the perimeter of the polygon.

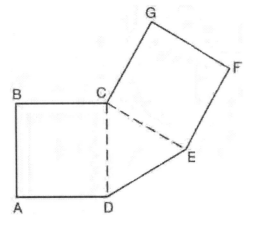

10. A garden, shaded in the diagram below, consists of four quarter-circles of equal size inside a square. The landscaper decides to put a fence around both the inside and the outside of the garden.

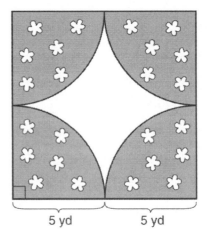

How much fencing, in yards, will the landscaper need for the fence?

11. In the figure below, arc *SBT* is one quarter of a circle with center *R* and radius 6. If the length plus the width of rectangle *ABCR* is 8, find the perimeter of the shaded region.
Hint: The diagonals of a rectangle are congruent to each other.

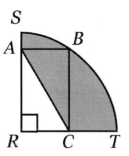

1.4 Area

MODEL PROBLEM 1: AREA OF A COMPOSITE FIGURE

The "key" in a regulation size basketball court is made up of a rectangle and semicircle. The length of the rectangle measures 19 feet from the baseline to the free throw line, and the width measures 12 feet, as shown below. Find the exact area of the key, in terms of π.

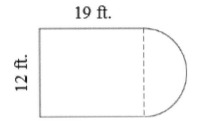

Solution:
(A) $A = lw + \frac{1}{2}\pi r^2$

(B) $= (19)(12) + \frac{1}{2}\pi(6)^2$

(C) $= 228 + 18\pi$ square feet

Explanation of steps:
(A) Write an equation that adds the areas of the parts that make up the composite figure *[rectangle and half of a circle].*
(B) Substitute the given values for the variables *[19, 12, and 6, for l, w and r].*
(C) Simplify. *[Write the answer in terms of π as the directions specify.]*

PRACTICE PROBLEMS

1. In the figure below, *ABCD* is a square and semicircle *O* has a radius of 6. What is the exact area of the figure?

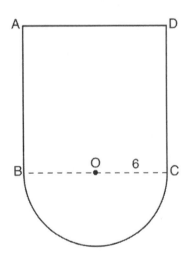

Basic Geometry 1.4 Area

2. The figure below consists of two semicircles connected to a 10 foot by 20 foot rectangle. Find the exact area of the composite figure.

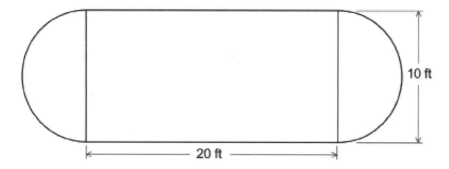

3. The figure below consists of three-quarters of a circle and a square with two of its sides formed by the circle's radii. If the radius of the circle is 4, find the exact area of the figure.

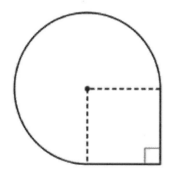

Basic Geometry
1.4 Area

4. The figure below consists of two semicircles and a square. The length of each side of the square region is represented by $2x$.

   Express the exact area of the figure in simplest terms of x.

   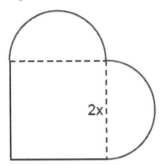

MODEL PROBLEM 2:   AREA OF A SHADED REGION

Find the area of the shaded region.

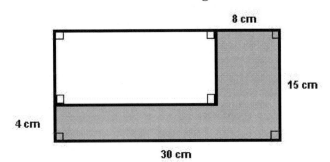

Solution:
(A) $A_{shaded} = A_{large} - A_{small}$
(B) $= (30)(15) - (22)(11)$
(C) $= 208$ sq. cm

Explanation of steps:
(A) Write an equation that subtracts the unshaded area *[the smaller rectangle]* from the total area *[the larger rectangle]* to find the shaded area.
(B) Calculate each area *[the sides of the larger rectangle, 30 and 15, are given, but the lengths of the sides of the smaller rectangle need to be calculated as 30 – 8 and 15 – 4]*.
(C) Simplify.

Practice Problems

5. A circle with radius 4 is inscribed inside a square as shown. Find the exact area of the shaded region.

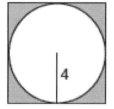

6. In the diagram below, a circle is inscribed inside a square. The square has an area of 36. What is the exact area of the circle?

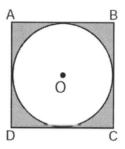

7. The figure below consists of a square and a semicircle. If the length of a side of the square is 6, what is the exact area of the shaded region?

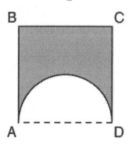

8. A circle and two semicircles are inscribed inside a 20 ft. by 10 ft. rectangle as shown. Find the exact area of the shaded region.

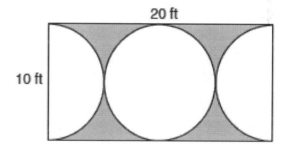

9. The diagram shows an isosceles right triangle inside a quarter-circle with a radius of 4 cm. Find the exact area of the shaded region.

10. The diagram shows a triangle inside a rectangle with dimensions as given. Find the area of the shaded region.

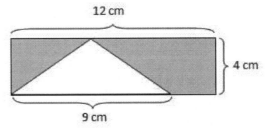

11. A designer created the logo shown below. The logo consists of a square and four quarter-circles of equal size. Find the exact area of the shaded region.

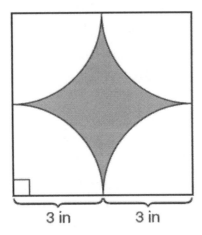

12. The Seedy Landscaping Company has been asked to lay grass sod in a rectangular area surrounding a circular fountain with a diameter of 8 feet. The area where the grass sod is to be laid is shaded in the diagram below. Find the shaded area, to the *nearest square foot*.

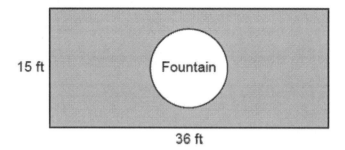

Find the cost, *in dollars*, of laying sod in the shaded area if the Seedy Landscaping Company charges $1.95 per square foot.

13. A rectangular garden is going to be planted in a person's rectangular backyard, as shown in the diagram below. Some dimensions of the backyard and the width of the garden are given. Find the area of the garden to the *nearest square foot.*

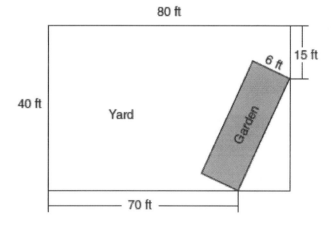

Model Problem 3: Area Density

The writeable area of a standard CD lies between two concentric circles with diameters of 116mm (the outer circle) and 50mm (the inner hole). The CD holds approximately 0.08135 mebibytes of data per square mm of writeable surface. How many total mebibytes can the CD hold, to the *nearest whole*?

Solution:
(A) $A = \pi(58)^2 - \pi(25)^2 = 2{,}739\pi \approx 8{,}604.82$ mm²
(B) $0.08135 = \dfrac{x}{8604.82}$
(C) $x = 0.08135 \times 8{,}604.82 \approx 700$ mebibytes

Explanation of steps:
(A) To find the shaded *[writeable]* area, subtract the area of the inner circle from the area of the outer circle. For each circle, the radius is half the diameter *[58 and 25]*. Substitute for r in the Area formula $A = \pi r^2$ to find the area of each circle and then find the difference.
(B) $density = \dfrac{total\ units}{area}$, so substitute for the known values *[the density is measured in quantity per square unit, so 0.08135 is the density]*.
(C) Solve.

Practice Problems

14. There are about 27,785 people per square mile of land in New York City. If the land area of the city is 303 square miles, what is the city's approximate population, to the *nearest thousand*?

15. A rectangular plantation measuring 2000 by 500 feet contains 6,887 apple trees. Using the following conversion equation, what is the density of apple trees per acre, to the *nearest whole*?

    1 acre = 43,560 square feet

Basic Geometry — 1.4 Area

16. In Manhattan, there are approximately 14.7 Starbucks coffee shops per 100,000 people.

   a) If the population of Manhattan is approximately 1,632,000, how many Starbucks coffee shops are there? Round to the *nearest whole*.

   b) Manhattan has a land area of approximately 22.8 square miles. How many Starbucks are there per square mile, to the *nearest tenth*?

Chapter 2. Coordinate Geometry

2.1 Forms of Linear Equations

MODEL PROBLEM

Convert the equation $y = \frac{5}{6}x + \frac{7}{4}$ to standard form.

Solution:
(A) $12y = 12\left[\frac{5}{6}x + \frac{7}{4}\right]$
$12y = 10x + 21$
(B) $-10x + 12y = 21$
(C) $10x - 12y = -21$

Explanation of steps:
(A) If m or b are fractions, multiply both sides by their least common denominator [LCD of 6 and 4 is 12].
(B) Subtract the x term from both sides.
(C) If this results in a negative coefficient of x, multiply both sides by -1. [Negate all terms.]

PRACTICE PROBLEMS

1. Determine whether the point $(-2,3)$ lies on the line whose equation is $y = 3x + 15$.

2. Write an equation of the line whose slope is -4 and y-intercept is 5.

3. Write, in point-slope form, an equation of the line that passes through the point $(1, -2)$ with a slope of -3.

4. Write, in point-slope form, an equation of the line that passes through the points $(-2, -3)$ and $(5, -5)$.

Coordinate Geometry   2.1 Forms of Linear Equations

5. Write, in point-slope form, an equation of the line that passes through the points (1,3) and (8,5).

6. Write, in point-slope form, an equation of the line that passes through the points (5,4) and (−5,0).

7. Rewrite the equation $y = 2x - 5$ in standard form.

8. Rewrite the equation $y = \frac{3}{4}x + \frac{1}{2}$ in standard form.

2.2 Parallel and Perpendicular Lines

MODEL PROBLEM

The equations of two distinct lines are $y = 3x - 6$ and $2y = 3x + 6$. Are the lines parallel?

Solution:
(A) For $y = 3x - 6$, the slope $m = 3$.
Solving $2y = 3x + 6$ for y:
$$\frac{2y}{2} = \frac{3x + 6}{2}$$
$y = \frac{3}{2}x + 3$, so the slope $m = \frac{3}{2}$.
(B) The lines are *not* parallel because the slopes are not equal.

Explanation of steps:
(A) Write each equation in slope-intercept form to determine the slope of each line.
[The first equation is already in slope-intercept form, $y = mx + b$, so the slope $m = 3$. The second equation needed to be transformed, showing a slope of $\frac{3}{2}$.]
(B) If the slopes are equal, the lines are parallel.
[These slopes are 3 and $\frac{3}{2}$, so they are not parallel.]

PRACTICE PROBLEMS

1. Which equation below represents a line that is parallel to the line, $y = -x + 4$?

   (1) $2y + 2x = 6$
   (2) $2y - x = 6$

2. Which equation below represents a line that is parallel to the line, $4x + 6y = 5$?

   (1) $-3y = 2x + 5$
   (2) $-6y + 4x = 5$

3. Which equation below represents a line that is perpendicular to the line, $y = 2x - 7$?

   (1) $y = 2x + \frac{1}{7}$
   (2) $y = -\frac{1}{2}x + 1$

4. Which equation below represents a line that is perpendicular to the line, $y = -5x + 2$?

   (1) $x - 5y = 25$
   (2) $5x - y = 5$

5. Line ℓ has an equation of $y = -2x - 5$. Write the equation of a line that is parallel to line ℓ but has a y-intercept of 2.

6. Line ℓ has an equation of $y = \frac{1}{2}x + 2$. Write an equation of a line perpendicular to line ℓ and passing through the origin.

7. Write an equation of the line passing through the point $(4, -1)$ and parallel to the line whose equation is $2y - x = 8$.

8. What is an equation of the line that passes through $(-9, 12)$ and is perpendicular to the line whose equation is $y = \frac{1}{3}x + 6$?

9. What is an equation of the line that passes through the point (2,4) and is perpendicular to the line whose equation is $3y = 6x + 3$?

10. Determine whether the lines represented by the equations $y = 2x + 3$ and $2y + x = 6$ are parallel, perpendicular, or neither. Justify your response.

11. Two lines are represented by the equations $x + 2y = 4$ and $4y - 2x = 12$. Determine whether these lines are parallel, perpendicular, or neither. Justify your answer.

12. The slope of $\overline{QR}$ is $\frac{x-1}{4}$ and the slope of $\overline{ST}$ is $\frac{8}{3}$. If $\overline{QR} \perp \overline{ST}$, determine and state the value of x.

2.3 Distance Formula

MODEL PROBLEM

Find the distance between the point $A(-2,-4)$ and the line $y = -\frac{1}{3}x + 2$.

Solution:
(A) For $y = -\frac{1}{3}x + 2$, $m = -\frac{1}{3}$, so $m_\perp = 3$
(B) Equation of the perpendicular line through $A(-2,-4)$ is $y + 4 = 3(x + 2)$,
or in slope-intercept form, $y = 3x + 2$
(C) $-\frac{1}{3}x + 2 = 3x + 2$, or $x = 0$
$y = 3(0) + 2$, or $y = 2$
So, intersection of lines is at $B(0,2)$
(D) $AB = \sqrt{(0-(-2))^2 + (2-(-4))^2}$
$= \sqrt{2^2 + 6^2} = \sqrt{40} = 2\sqrt{10}$

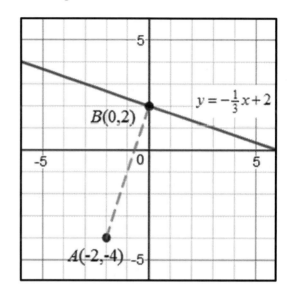

Explanation of steps:
(A) Find the slope of the line segment between the point and the line that is perpendicular to the line. It is the opposite reciprocal of the slope of the line.
[The opposite reciprocal of $-\frac{1}{3}$ is 3.]
(B) Write the equation of the line segment in point-slope form, using the given point and the slope from part (A).
(C) To find the point of intersection, solve the system of equation by substitution.
[Given $y = -\frac{1}{3}x + 2$ and $y = 3x + 2$, substitute for y in the second equation, and solve to get x = 0. Then substitute for x in the second equation and solve to get y = 2. These coordinates give us the point of intersection, (0,2).]
(D) Use the distance formula to find the distance between the two points
[$A(-2,-4)$ and $B(0,2)$].

Coordinate Geometry — 2.3 Distance Formula

PRACTICE PROBLEMS

1. Find the distance between the points $(-2, 3)$ and $(6, -3)$.

2. Find the distance between the points $(3, 5)$ and $(8, 10)$, in simplest radical form.

3. What is the length of the line segment whose endpoints are $(-1, 9)$ and $(7, 4)$?

4. What is the length of the line segment whose endpoints are $(5, 3)$ and $(1, 6)$?

5. The endpoints of $\overline{AB}$ are $A(3, -4)$ and $B(7, 2)$. Determine and state the length of $\overline{AB}$ in simplest radical form.

6. The coordinates of the endpoints of $\overline{CD}$ are $C(3, 8)$ and $D(6, -1)$. Find the length of $\overline{CD}$ in simplest radical form.

7. The endpoints of $\overline{PQ}$ are $P(-3,1)$ and $Q(4,25)$. Find the length of $\overline{PQ}$.

8. What is the length, in simplest radical form, of the line segment joining the points $(-4,2)$ and $(146,52)$?

9. Find the distance between the point $(6,-2)$ and the line $y = \frac{1}{5}x + 2$.

Coordinate Geometry 2.4 Midpoint Formula

2.4 Midpoint Formula

MODEL PROBLEM

The endpoints of $\overline{AB}$ are $A(-2,5)$ and $B(4,11)$. Find the coordinates of the midpoint of $\overline{AB}$.

Solution:
$$\left(\frac{x_1 + x_2}{2}, \frac{y_1 + y_2}{2}\right) = \left(\frac{(-2) + 4}{2}, \frac{5 + 11}{2}\right) = (1,8)$$

Explanation:
Substitute the coordinates (x_1, y_1) and (x_2, y_2) [$(-2,5)$ and $(4,11)$] into the midpoint formula, and simplify.

PRACTICE PROBLEMS

1. Find the midpoint of the segment whose endpoints are $(-2,3)$ and $(6,-3)$.	2. The diameter of a circle has endpoints of $(3,5)$ and $(8,10)$. Find the coordinates of the center of the circle.

3. Quadrilateral $ABCD$ has vertices $A(-5,1)$, $B(6,-1)$, $C(3,5)$, and $D(-2,7)$. What are the coordinates of the midpoint of diagonal $\overline{AC}$?

4. The midpoint of $\overline{AB}$ is $M(3,5)$. Given $A(1,-5)$, find the coordinates of B.

5. In triangle ABC, median $\overline{AM}$ is drawn. Given $A(3,4)$, $B(4,-3)$, and $M(0,-2)$, find the coordinates of C.

6. In the diagram of circle C, $\overline{QR}$ is a diameter, and the coordinates of points Q and C are given. Find the coordinates of point R.

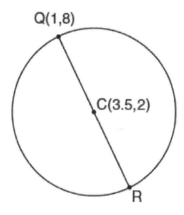

7. In a circle whose center is $(2,3)$, one endpoint of a diameter is $(-1,5)$. Find the coordinates of the other endpoint of that diameter.

2.5 Perpendicular Bisectors

Model Problem

Write an equation for the perpendicular bisector of $\overline{PQ}$, shown in the graph to the right.

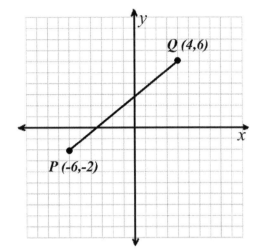

Solution:

(A) Midpoint is $\left(\dfrac{-6+4}{2}, \dfrac{-2+6}{2}\right) = (-1, 2)$.

(B) Slope is $\dfrac{6-(-2)}{4-(-6)} = \dfrac{8}{10} = \dfrac{4}{5}$.

(C) Slope of perpendicular line is $-\dfrac{5}{4}$. Equation of perpendicular bisector is $y - 2 = -\dfrac{5}{4}(x+1)$.

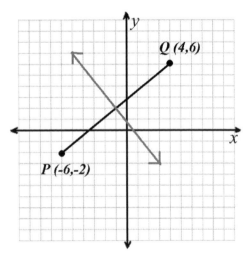

Explanation of steps:

(A) Find the coordinates of the midpoint *[using the midpoint formula]*.
(B) Find the slope of the line segment *[using the slope formula]*.
(C) Write an equation of the line that passes through the midpoint and has an opposite reciprocal slope. We can use the point-slope form, $y - y_1 = m(x - x_1)$. *[Substitute the midpoint coordinates $(-1, 2)$ for x_1 and y_1 and the slope for m.]*

Coordinate Geometry — 2.5 Perpendicular Bisectors

Practice Problems

1. Write an equation of the perpendicular bisector of the segment whose endpoints are (3,5) and (9,17).

2. Write an equation of the perpendicular bisector of the segment whose endpoints are (−2,3) and (6,−3).

3. Given $A(2,6)$ and $B(8,12)$, write an equation of $\overleftrightarrow{CD}$, the perpendicular bisector of $\overline{AB}$. What is the y-intercept of $\overleftrightarrow{CD}$?

4. Write an equation of the perpendicular bisector of the segment whose endpoints are (−4,5) and (2,5).

5. Write an equation of the perpendicular bisector of $\overline{AB}$ whose endpoints are $A(4,2)$ and $B(8,6)$.

6. Write an equation of the perpendicular bisector of the line segment whose endpoints are $(-1,1)$ and $(7,-5)$.

7. Write an equation of the line that is the perpendicular bisector of the line segment having endpoints $(3,-1)$ and $(3,5)$.

8. The coordinates of the endpoints of $\overline{AB}$ are $A(0,0)$ and $B(6,0)$. Write an equation of the perpendicular bisector of $\overline{AB}$.

2.6 Directed Line Segments

Model Problem

Given points $A(3,4)$ and $B(6,10)$, find the coordinates of point P along the directed line segment AB so that the ratio of AP to PB is $3:2$.

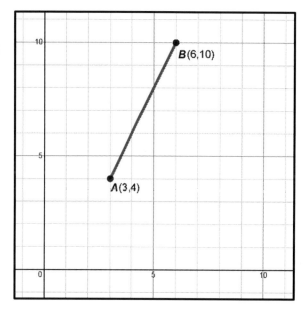

Solution:
(A) $AP = \frac{3}{5} AB$, so $k = \frac{3}{5}$.
(B) $P_x = A_x + k(B_x - A_x) = 3 + \frac{3}{5}(6-3) = 4.8$
(C) $P_y = A_y + k(B_y - A_y) = 4 + \frac{3}{5}(10-4) = 7.6$
(D) Point P is $(4.8, 7.6)$

Explanation of steps:
(A) Convert the ratio [$3:2$] to a fraction, k, representing part over whole [$\frac{3}{5}$].
(B) The x-coordinate of P will be the x-coordinate of A plus the fraction k of the *run* from A to B.
(C) The y-coordinate of P will be the y-coordinate of A plus the fraction k of the *rise* from A to B.
(D) State the coordinates of point P.

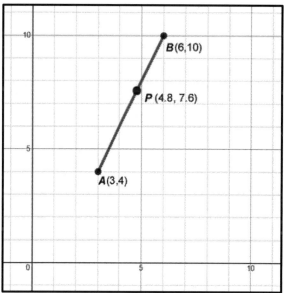

Coordinate Geometry — 2.6 Directed Line Segments

Practice Problems

1. What are the coordinates of the point P on the directed line segment from $A(3,5)$ to $B(9,17)$ that partitions the segment into the ratio of $5:1$.

2. What are the coordinates of the point S on the directed line segment from $R(-2,2)$ to $T(3,-8)$ that partitions the segment into the ratio of $3:2$.

3. What are the coordinates of the point P on the directed line segment from $L(-2,3)$ to $M(6,-3)$ such that $LP:PM$ is $3:5$.

4. What are the coordinates of the point G on the directed line segment from $F(1,-3)$ to $H(6,5)$ such that $FG:GH$ is $2:3$.

Chapter 3. Polygons in the Coordinate Plane

3.1 Triangles in the Coordinate Plane

Model Problem

Given $\triangle KLM$ with $K(1,2)$, $L(3,3)$, $M(2,y)$, and $\angle L$ is a right angle, find the value of y. Prove that $\angle K \cong \angle M$.

Solution:

(A) $m_{\overline{KL}} = \dfrac{3-2}{3-1} = \dfrac{1}{2}$

(B) $m_{\overline{LM}} = -2$

(C) $\dfrac{y-3}{2-3} = -2$

(D) $y - 3 = 2$
$y = 5$
$M(2,5)$

(E) $LM = \sqrt{(2-3)^2 + (5-3)^2} = \sqrt{5}$
$KL = \sqrt{(3-1)^2 + (3-2)^2} = \sqrt{5}$

(F) Since $\overline{LM} \cong \overline{KL}$ and the angles opposite congruent sides in a triangle are congruent, then $\angle K \cong \angle M$.

Explanation of steps:

(A) The legs of the right triangle *[$\overline{KL}$ and $\overline{LM}$]* are perpendicular. Find the slope of one leg *[the slope of $\overline{KL}$ is $\frac{1}{2}$]*.

(B) The slope of the legs are opposite reciprocals *[so, the slope of $\overline{LM}$ is –2]*.

(C) Substitute into the slope formula using the coordinates of the endpoints *[L and M]*.

(D) Solve for y. This gives us the coordinates of the third vertex *[M]*.

(E) To prove that two angles of a triangle are congruent, we can show that their opposite sides are congruent. Use the distance formula to find the lengths of the opposite sides. Congruent sides are equal in length.

(F) State the conclusion and reason.

Practice Problems

1. Determine whether △ABC is a right or oblique triangle. Justify your answer.

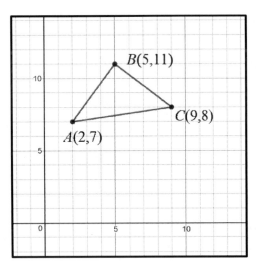

2. For △ABC shown in the previous question, determine whether the triangle is scalene, isosceles, or equilateral.

3. Given △DEF with $D(4,-2)$, $E(5,5)$, and $F(-1,3)$. Determine whether △DEF is scalene, isosceles, or equilateral.

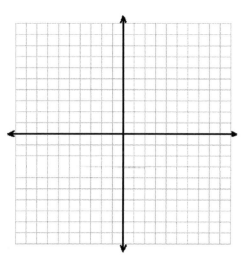

4. Right triangle JKL has a right angle at ∠K and vertices $J(-2,4)$, $K(6,6)$, and $L(x,-2)$. Find x.

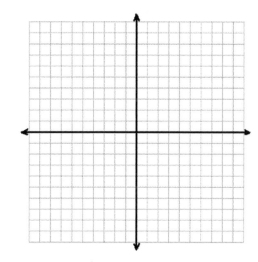

5. Find the area of triangle PQR with vertices $P(4,-2)$, $Q(-6,4)$, and $R(8,-2)$.

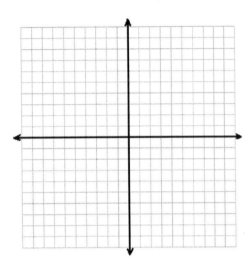

6. Given: $A(2,2), B(5,1), C(4,5), D(1,-4), E(4,-5), F(3,-1)$
   Prove: $\triangle ABC \cong \triangle DEF$

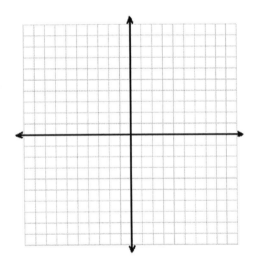

7. Given: $R(-1,7), S(3,-1)$, and $T(9,2)$
   Prove: $\triangle RST$ is a right triangle

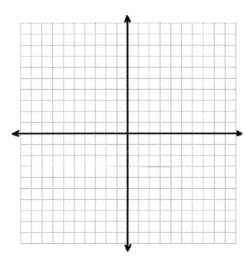

8. Given: $J(-4,1), E(-2,-3), N(2,-1)$
   Prove: $\triangle JEN$ is an isosceles right triangle.

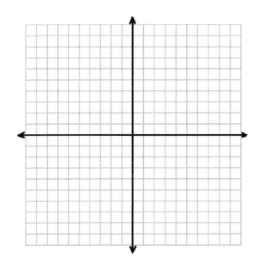

3.2 Quadrilaterals in the Coordinate Plane

Model Problem

Given: Parallelogram $ABCD$ with $A(-2,6), B(4,3), C(2,-1), D(-4,2)$.
Prove: $ABCD$ is a rectangle.

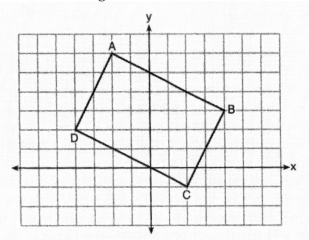

Solution:

(A)
Method 1: Show any angle is a right angle.
$m_{\overline{AB}} = \dfrac{3-6}{4+2} = -\dfrac{1}{2}$ $m_{\overline{BC}} = \dfrac{-1-3}{2-4} = 2$

The slopes of $\overline{AB}$ and $\overline{BC}$ are opposite reciprocals, so $\overline{AB} \perp \overline{BC}$ and $\angle B$ is a right angle. A parallelogram with a right angle is a rectangle, so $ABCD$ is a rectangle.

(B)
Method 2: Show diagonals are congruent.
$AC = \sqrt{(2+2)^2 + (-1-6)^2} = \sqrt{65}$
$BD = \sqrt{(-4-4)^2 + (2-3)^2} = \sqrt{65}$
Diagonals $\overline{AC}$ and $\overline{BD}$ have the same length, so $\overline{AC} \cong \overline{BD}$. A parallelogram with congruent diagonals is a rectangle, so $ABCD$ is a rectangle.

Explanation of steps:

We can use either one of the following methods to prove $ABCD$ is a rectangle:
(A) show that any angle is a right angle, or
(B) show that the diagonals are congruent.

(A) We can show an angle *[such as $\angle B$]* is a right angle by showing the sides that form the vertex *[$\overline{AB}$ and $\overline{BC}$]* are perpendicular. Use the slope formula to show their slopes are opposite reciprocals.

(B) We can show the diagonals *[$\overline{AC}$ and $\overline{BD}$]* are congruent by using the distance formula to find their lengths. Diagonals with equal lengths are congruent.

Polygons in the Coordinate Plane   *3.2 Quadrilaterals in the Coordinate Plane*

PRACTICE PROBLEMS

1. Parallelogram *ABCD* has vertices $A(1,3)$, $B(5,7)$, $C(10,7)$, and $D(6,3)$. Diagonals $\overline{AC}$ and $\overline{BD}$ intersect at *E*. What are the coordinates of point *E*?

2. Given: Quadrilateral ABCD with $A(-5,0), B(-1,-8), C(7,-4), D(3,4)$.
   Prove: *ABCD* is a rectangle.

3. Quadrilateral *ABCD* has vertices $A(-6,-3), B(1,0), C(4,7)$, and $D(-3,4)$. Classify *ABCD* using the most precise name.

4. Quadrilateral *ABCD* has vertices $A(-5,-6), B(2,0), C(11,9),$ and $D(4,3)$. Classify *ABCD* using the most precise name.

5. Quadrilateral *ABCD* has vertices $A(1,1), B(5,2), C(6,-2),$ and $D(2,-3)$. Classify *ABCD* using the most precise name.

6. The coordinates of two vertices of square *ABCD* are $A(2,1)$ and $B(4,4)$. Determine the slope of side $\overline{BC}$.

7. The vertices of square *PQRS* are $P(-4,0), Q(4,3), R(7,-5),$ and $S(-1,-8)$. Show that the diagonals of square *PQRS* are congruent perpendicular bisectors of each other.

8. Three of the vertices of parallelogram ABCD are A(0,0), B(5,2), and C(6,5). Find the coordinates of point D. Justify that the figure is a parallelogram.

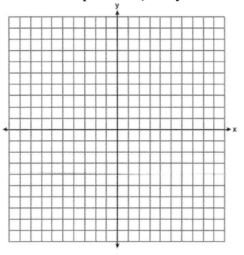

9. The vertices of quadrilateral ABCD are A(−1, −5), B(8,2), C(11,13), and D(2,6). Prove that quadrilateral ABCD is a rhombus.

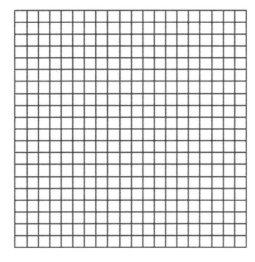

10. Given: $A(-2,2), B(6,5), C(4,0), D(-4,-3)$
    Prove: $ABCD$ is a parallelogram but *not* a rectangle.

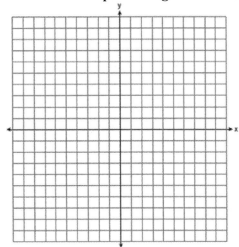

11. Quadrilateral $MATH$ has coordinates $M(1,1), A(-2,5), T(3,5),$ and $H(6,1)$. Prove that quadrilateral $MATH$ is a rhombus but *not* a square.

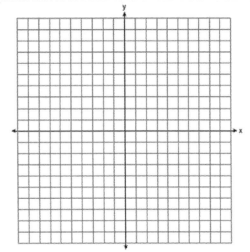

12. Given: Quadrilateral ABCD has vertices $A(-5,6), B(6,6), C(8,-3)$, and $D(-3,-3)$.
    Prove: Quadrilateral ABCD is a parallelogram but is *neither* a rhombus *nor* a rectangle.

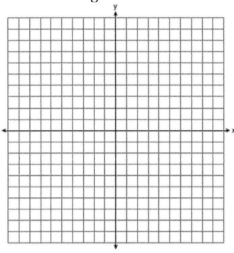

13. Quadrilateral ABCD with vertices $A(-7,4), B(-3,6), C(3,0)$, and $D(1,-8)$ is graphed on the set of axes below.

    Quadrilateral MNPQ is formed by joining M, N, P, and Q, the midpoints of $\overline{AB}, \overline{BC}, \overline{CD}$, and $\overline{AD}$, respectively. Prove that quadrilateral MNPQ is a parallelogram but is *not* a rhombus.

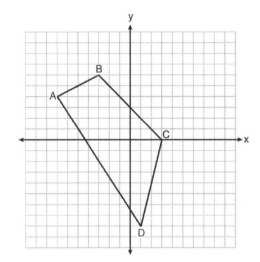

3.3 Perimeter and Area using Coordinates

MODEL PROBLEM 1: *USING THE DISTANCE FORMULA*

Find the perimeter and area of rectangle $ABCD$, whose vertices are points $A(-3,0)$, $B(3,2)$, $C(4,-1)$, and $D(-2,-3)$.

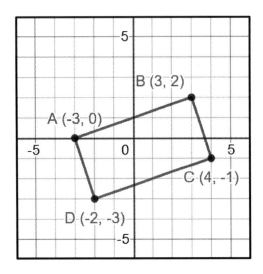

Solution:
(A) $AB = \sqrt{6^2 + 2^2} = \sqrt{40} = 2\sqrt{10}$ $CD = 2\sqrt{10}$
(B) $BC = \sqrt{1^2 + (-3)^2} = \sqrt{10}$ $AD = \sqrt{10}$
(C) Perimeter = $2\sqrt{10} + 2\sqrt{10} + \sqrt{10} + \sqrt{10} = 6\sqrt{10} \approx 19.0$
(D) Area = $2\sqrt{10} \cdot \sqrt{10} = 20$ square units

Explanation of steps:
(A) Use the distance formula to find the base *[AB]*. Since it is a rectangle, the opposite side *[CD]* is the same length.
(B) Use the distance formula to find the height *[BC]*. In a rectangle, its opposite side *[AD]* is also equal in length.
(C) Add the two bases and two heights to find the perimeter.
(D) Multiply the base and height to find the area. *[We are able to calculate $A = bh$ directly.]*

Polygons in the Coordinate Plane — 3.3 Perimeter and Area using Coordinates

Practice Problems

1. Find the perimeter of △ABC below.

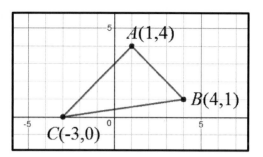

2. Triangle ABC has coordinates $A(-6,2)$, $B(-3,6)$, and $C(5,0)$. Find the perimeter of the triangle. Express your answer in simplest radical form.

3. Find the area of △ABC below.

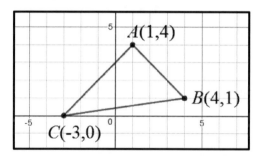

4. Parallelogram *EFGH* has vertices $E(3,6)$, $F(6,10)$, $G(18,5)$, and $H(15,1)$. Find its perimeter and area.

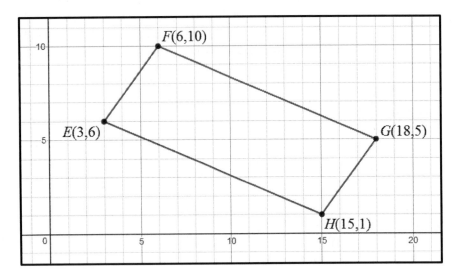

Polygons in the Coordinate Plane — 3.3 Perimeter and Area using Coordinates

MODEL PROBLEM 2: *USING THE SHOELACE METHOD*

Find the area of the quadrilateral *ABCD*, whose vertices are points $A(-3,0), B(2,4), C(3,1)$, and $D(-4,-3)$, using the shoelace method.

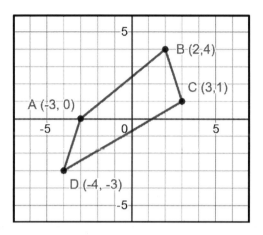

Solution:

(A) vertex	x	y	(B) upper	lower	(C) difference
A	−3	0	−12	0	−12
B	2	4	2	12	−10
C	3	1	−9	−4	−5
D	−4	−3	0	9	−9
A	−3	0			

(D) $A = |(-12) + (-10) + (-5) + (-9)| \div 2 = |-36| \div 2 = 18$ square units

Explanation of steps:

(A) Create a table listing the coordinates of the vertices of the polygon in clockwise order, ending with the starting vertex.

(B) Find the "upper lace" products
$[-3 \times 4 = -12; 2 \times 1 = 2; 3 \times -3 = -9; -4 \times 0 = 0]$
and the "lower lace" products
$[0 \times 2 = 0; 4 \times 3 = 12; 1 \times -4 = -4; -3 \times -3 = 9]$.

(C) Calculate each difference, *upper − lower*
$[(-12) - 0 = -12; 2 - 12 = -10; etc.]$.

(D) Add the differences, take the absolute value of the sum, and divide by 2.

Practice Problems

5. Parallelogram *KLMN* has vertices $K(-7,-7), L(-5,2), M(3,6),$ and $N(1,-3)$. Find its area using the shoelace method.

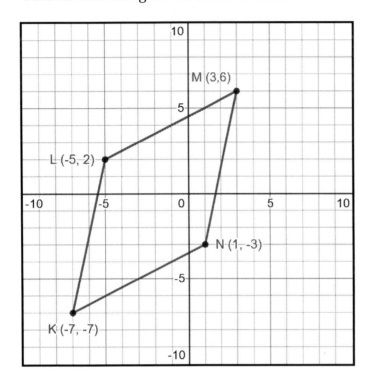

Chapter 4. Rigid Motions

4.1 Translations

Model Problem

A translation maps $P(3,-2)$ to $P'(1,2)$. Under the same translation, find the coordinates of Q', the image of $Q(-3,2)$.

Solution:
(A) Translation is $T_{-2,4}$.
(B) $Q'(-5,6)$.

Explanation of steps:
(A) Find the change in x and y by subtracting the coordinates of a point in the pre-image from its image.
[Subtracting the coordinates of P from P' gives us $1-3=-2$ and $2-(-2)=4$]
(B) Determine the image of the other point.
[Add –2 and 4 to the coordinates of Q to find Q']

Practice Problems

1. What is the image of point (2,4) under the translation $T_{-6,1}$?	2. What is the image of the point $(-5,2)$ under the translation $T_{3,-4}$?
3. What is the image of (x,y) after a translation of 3 units right and 7 units down?	4. What is the image of point $(3,-5)$ under the translation that shifts (x,y) to $(x-1, y-3)$?

5. What is the image of point $(-3,4)$ under the translation that shifts (x,y) to $(x-3, y+2)$?

6. What are the coordinates of P', the image of point $P(x,y)$ after translation $T_{4,4}$?

7. A translation moves $P(3,5)$ to $P'(6,1)$. What are the coordinates of the image of point $(-3,-5)$ under the same translation?

8. A translation maps $P(3,-2)$ onto $P'(5,0)$. Find the coordinates of the image of $Q(4,-6)$ under the same translation.

9. The vertices of a rectangle are $R(-5,-5)$, $S(-1,-5)$, $T(-1,1)$ and $U(-5,1)$. After a translation, R', the image of R, is the point $(3,0)$. Describe the translation and state the coordinates of U', the image of U.

10. Triangle ABC has vertices $A(1,3)$, $B(0,1)$, and $C(4,0)$. Under a translation, A', the image of A, is $(4,4)$. Describe the translation and state the coordinates of point C', the image of C?

11. Under the transformation $T_{2,-1}$ on point A, the image is point $A'(-3,4)$. What are the coordinates of A?

12. The image of △ABC under a translation is △A'B'C'. Under this translation, $B(3,-2)$ maps onto $B'(1,-1)$. Using this translation, the coordinates of image A' are $(-2,2)$. What are the coordinates of point A?

13. Rectangle A'B'C'D' is the image of a translation of rectangle ABDC. The table of translations is shown below. Find the coordinates of points B and D'.

Rectangle ABCD	Rectangle A'B'C'D'
A(2,4)	A'(3,1)
B(,)	B'(-5,1)
C(2,-1)	C'(3,-4)
D(-6,-1)	D'(,)

14. On the grid below, graph and label △ K'L'M', the image of △ KLM after a translation of $T_{6,-4}$.

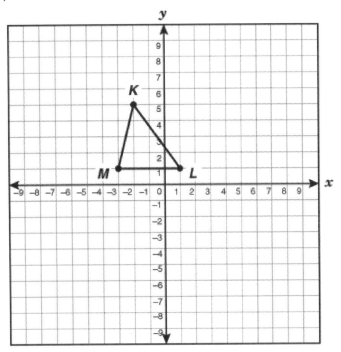

15. Triangle TAP has coordinates $T(-1,4)$, $A(2,4)$, and $P(2,0)$. On the set of axes below, graph and label △ T'A'P', the image of △ TAP after the translation $(x, y) \rightarrow (x - 5, y - 1)$.

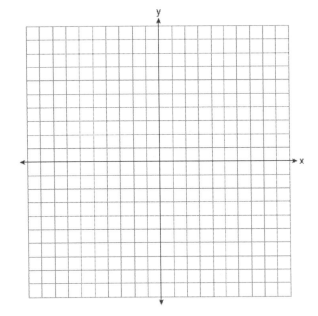

4.2 Line Reflections

Model Problem

Graph and label △ $G'L'Q'$, the image of △ GLQ under the transformation r_{y-axis}.

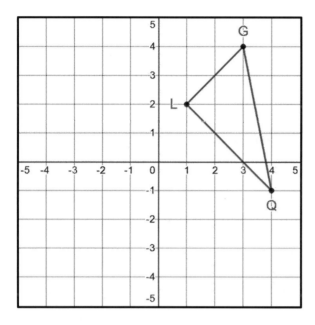

Solution:

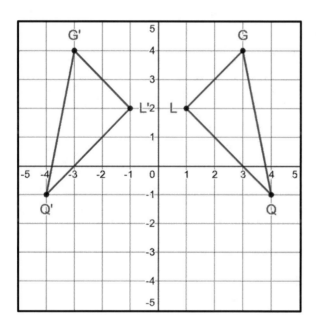

Explanation of steps:
When reflecting over the y-axis, use the rule, $r_{y-axis}: (x, y) \rightarrow (-x, y)$. That is, negate the x-coordinate of each point but keep the same y-coordinates.
[$L(1,2) \rightarrow L'(-1,2)$, $G(3,4) \rightarrow G'(-3,4)$, and $Q(4,-1) \rightarrow Q'(-4,-1)$]

Rigid Motions   4.2 Line Reflections

Practice Problems

1. What is the image of point $(2, -3)$ after it is reflected over the *x*-axis?

2. If $P(-4, -1)$ is reflected in the *x*-axis, what are the coordinates of P', the image of P?

3. When the point $(2, -5)$ is reflected in the *x*-axis, what are the coordinates of its image?

4. What is the image of point $(3, 4)$ when reflected in the *y*-axis?

5. If $M(-2, 8)$ is reflected in the *y*-axis, what are the coordinates of M', the image of M?

6. What are the coordinates of point P, the image of point $(3, -4)$, after a reflection in the line $y = x$?

7. What is the image of point $(5, -2)$ under the transformation $r_{y=x}$?

8. The endpoints of $\overline{AB}$ are $A(0,2)$ and $B(4,6)$. State the coordinates of A' and B', the images of A and B after $\overline{AB}$ is reflected in the *x*-axis.

9. Graph and label △A'B'C', the image of △ABC after a reflection over the line $y = -1$.

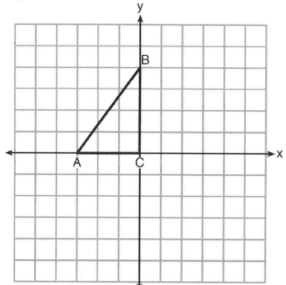

10. Graph and label △A'B'C', the image of △ABC after a reflection over the line $y = x$.

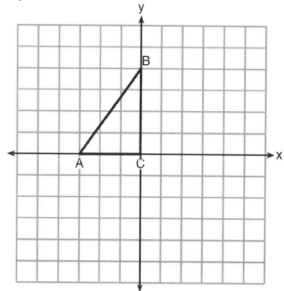

11. Triangle XYZ, shown in the diagram below, is reflected over the line $x = 2$. State the coordinates of △X'Y'Z', the image of △XYZ.

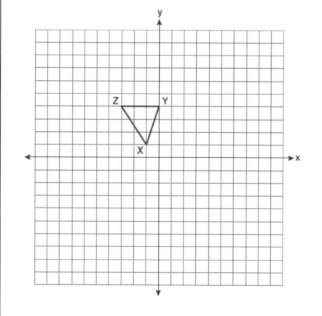

12. △SUN has coordinates $S(0,6)$, $U(3,5)$, and $N(3,0)$. On the grid below, draw and label △SUN. Then, graph and state the coordinates of △S'U'N', the image of △SUN after a reflection in the y-axis.

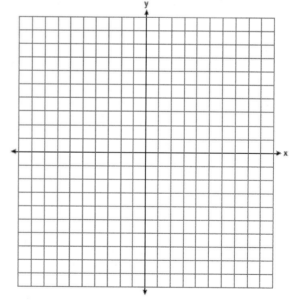

13. On the accompanying set of axes, draw the reflection of ABCD in the y-axis. Label and state the coordinates of the reflected figure.

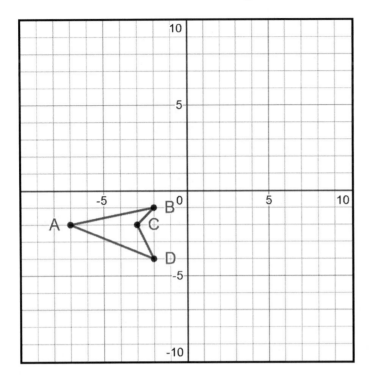

14. The coordinates of a square are $A(2,6), B(7,8), C(9,3)$, and $D(4,1)$.

   a) On the grid below, sketch and label ABCD and its image A'B'C'D', which is the reflection of ABCD over the y-axis.
   b) Find the number of square units in the area of ABCD.

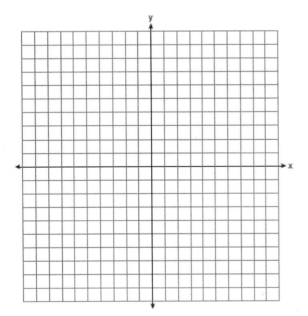

15. Find the image of $P(-2,2)$ after a reflection over the line $y = \frac{1}{2}x - 2$.

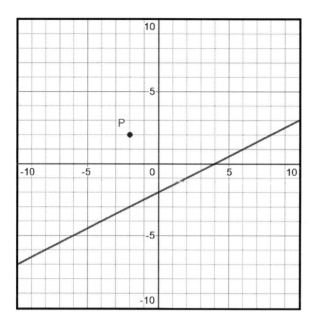

16. Find the image of $P(-2,2)$ after a reflection over the line $y = -x + 3$.

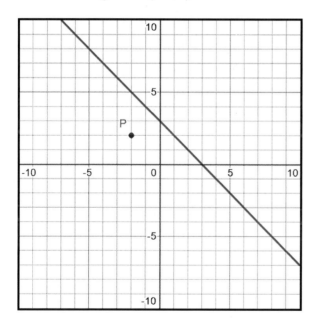

4.3 Rotations

MODEL PROBLEM

Which figure represents a rotation of triangle *A*?

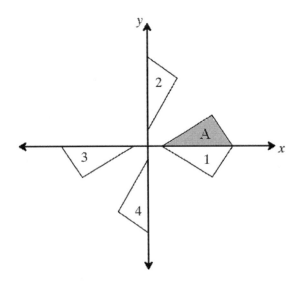

Solution:

Figure 3

Explanation:

A rotation will always preserve orientation. *[Figure 3 is the image after $R_{(0,0),180°}$. It is the only figure among the four that has the same orientation as triangle A.]*

PRACTICE PROBLEMS

1. In the graph below, the solid rectangle is the image of the dashed rectangle under a rotation

    (1) 180° about the origin.

    (2) 90° counterclockwise about the origin.

    (3) 90° clockwise about the origin.

    (4) 270° counterclockwise about the origin.

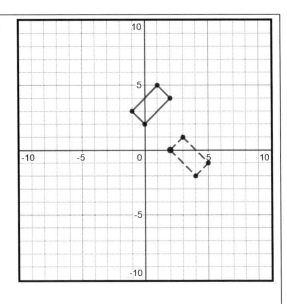

2. The diagram shows the starting position of the spinner on a board game. How does this spinner appear after a 270° counterclockwise rotation about point *P*?

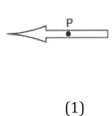

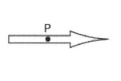

(1) (2) (3) (4)

3. Which graph represents the image of trapezoid *ABCD* after a rotation of 180° around vertex *D*?

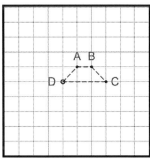

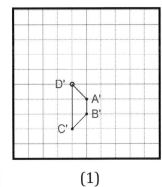 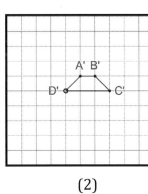 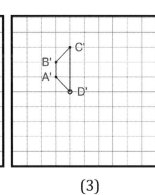 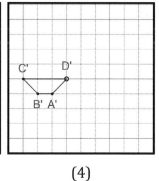

(1) (2) (3) (4)

4. What are the coordinates of M', the image of $M(2,4)$, after a *counterclockwise* rotation of 90° about the origin?	5. What are the coordinates of the image of the point $(-2,5)$ after a *clockwise* rotation of 90° about the origin?
6. The point $(-2,1)$ is rotated 180° about the origin in a *clockwise* direction. What are the coordinates of its image?	7. Rectangle $ABCD$ has vertices $A(0,-4)$, $B(4,-2)$, $C(5,-4)$, and $D(1,-6)$. Find the coordinates of the vertices of $A'B'C'D'$, the image of $ABCD$ after a rotation of 180° about the origin.

8. The coordinates of the vertices of △ABC are $A(1,2)$, $B(-4,3)$, and $C(-3,-5)$. State the coordinates of △A'B'C', the image of △ABC after a rotation of 90° about the origin.

9. What are the coordinates of the image of the point $P(-2,5)$ after a 90° rotation about the point $C(2,3)$?

10. What are the coordinates of the image of the point $P(3,-2)$ after a 180° rotation about the point $C(2,-3)$?

11. The coordinates of the vertices of △ABC are $A(1,2)$, $B(-4,3)$, and $C(-3,-5)$. State the coordinates of △A'B'C', the image of △ABC after a rotation of 90° about the point $P(2,-1)$.

4.4 Point Reflections

Model Problem

Given $\triangle ABC$ with coordinates $A(-1,4)$, $B(4,-4)$, and $C(-1,-6)$. Find the coordinates of the vertices of $\triangle A'B'C'$, the image of $\triangle ABC$ after a reflection in the point $P(3,4)$.

Solution:
$A(-1,4)$ maps to $(2 \cdot 3 - (-1), 2 \cdot 4 - 4)$, which is $A'(7,4)$.
$B(4,-4)$ maps to $(2 \cdot 3 - 4, 2 \cdot 4 - (-4))$, which is $B'(2,12)$.
$C(-1,-6)$ maps to $(2 \cdot 3 - (-1), 2 \cdot 4 - (-6))$, which is $C'(7,14)$.

Explanation of steps:
To find the image of each point,
(1) calculate the x-value by doubling the x-value of the point of reflection *[2 · 3]* and subtracting the x-value of the original point, and
(2) calculate the y-value by doubling the y-value of the point of reflection *[2 · 4]* and subtracting the y-value of the original point.

Practice Problems

1. Find the coordinates of the point K', the image of $K(4,-7)$ after a reflection through the origin.	2. What is the image of point $(-3,-1)$ under a reflection in the origin?
3. The image of $\overline{RS}$ after a reflection through the origin is $\overline{R'S'}$. If the coordinates of the endpoints of $\overline{RS}$ are $R(2,-3)$ and $S(5,1)$, what are the coordinates of R' and S'?	4. Find the coordinates of the point N', the image of $N(5,3)$ after a reflection through the point $P(1,-1)$.

5. Point M' is the image of point M after a reflection in point R. If the length of $\overline{MR}$ is 6 units, find the length of $\overline{MM'}$.

6. Given $B(3,6)$ and $B'(7,-2)$, where B' is the image of B after a reflection in the point P. Find the coordinates of P.

7. Graph the image of $\overline{AB}$ after a reflection through the origin.

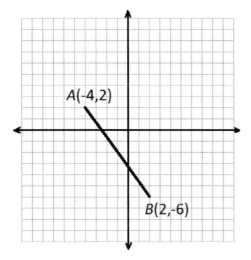

8. The graph of $\triangle DEF$ is shown below, with vertices $D(-8,5)$, $E(-6,8)$, and $F(-5,2)$. Graph $\triangle D'E'F'$, the image of $\triangle DEF$ after a reflection through $(0,3)$.

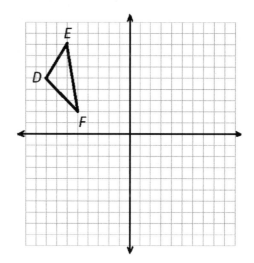

4.5 Carry a Polygon onto Itself

MODEL PROBLEM

To the right is an image of a nautical steering wheel. If the wheel is turned, what is the minimum number of degrees that it would need to rotate in order to carry onto itself?

Solution:

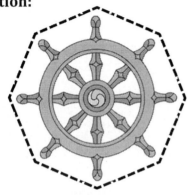

$$\frac{360°}{8} = 45°$$

Explanation of steps:
For a regular polygon, divide 360° by the number of sides. *[By connecting the tips of the spokes of the wheel, we form a regular octagon, so there are 8 sides.]*

PRACTICE PROBLEMS

1. Which reflection would carry this regular pentagon onto itself?

   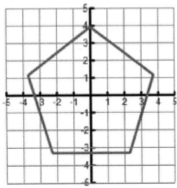

   (1) reflection over $y = 0$
   (2) reflection over $y = 1$
   (3) reflection over $x = 0$
   (4) reflection over $y = x$

2. What angle of rotation around $(-1, 2)$ would carry this rectangle onto itself?

   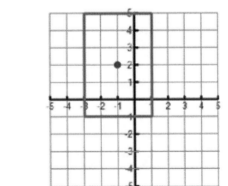

3. What is the minimum number of degrees that the airplane propeller shown below would need to rotate in order to carry onto itself?

4. The bottom car of the Ferris wheel below is positioned for riders to board. How many degrees must the wheel turn for riders to board an adjacent car?

Chapter 5. Dilations

5.1 Dilations of Line Segments

Model Problem

$\overline{M'N'}$ is the image of $\overline{MN}$ under a dilation with the origin as the center and a scale factor of 2.5. Given $M(4,2)$ and $N(-2,0)$, find the coordinates of M' and N'.

Solution:
$M(4,2) \rightarrow M'(10,5)$ and $N(-2,0) \rightarrow N'(-5,0)$

Explanation of steps:
Use $(x,y) \rightarrow (rx, ry)$ to multiply the coordinates of each endpoint by the scale factor $[r = 2.5]$.

Practice Problems

1. Which graph shows a dilation that has a scale factor of $\frac{1}{2}$ and center at the origin?

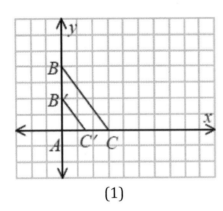

(1)

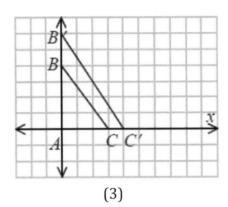

(3)

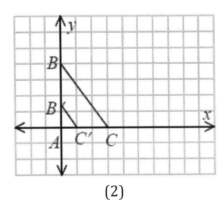

(2)

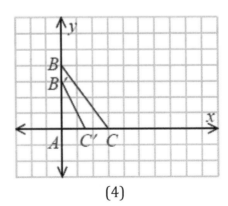

(4)

2. What is the image of $(3,-2)$ under the dilation $D_{(0,0),5}$?	3. What is the image of point $A(1,3)$ after a dilation with the center at the origin and a scale factor of 4?
4. Find the image of $(3,-2)$ for a dilation centered at the origin with a scale factor of $\frac{1}{2}$.	5. The image of point A after a dilation centered at the origin with a scale factor of 3 is $(6,15)$. What was the original location of point A?
6. Under a dilation where the center of dilation is the origin, the image of $A(-2,-3)$ is $A'(-6,-9)$. What are the coordinates of B', the image of $B(4,0)$ under the same dilation?	7. Graph $\overline{JK}$ with $J(2,3)$ and $K(-4,4)$. Then graph its dilation centered at the origin with a scale factor of 2. 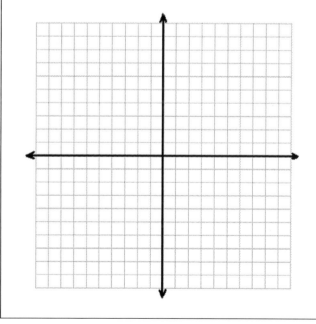

8. $\overline{AB}$ has endpoints of $A(1,6)$ and $B(6,6)$. Graph and state the endpoints of $\overline{A'B'}$, the image of $\overline{AB}$ after a dilation centered at point $P(3,4)$ with a scale factor of 2.

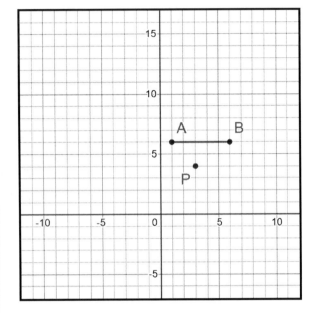

9. $\overline{CD}$ has endpoints of $C(0,6)$ and $D(2,0)$. Graph and state the endpoints of $\overline{C'D'}$, the image of $\overline{CD}$ after a dilation centered at point $P(-6,2)$ with a scale factor of $\frac{1}{2}$.

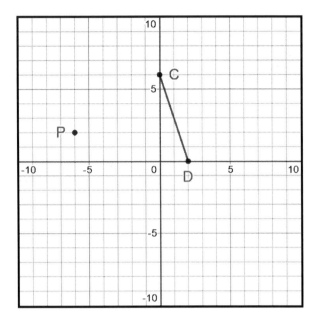

5.2 Dilations of Polygons

Model Problem

Graph the dilation of quadrilateral *MNOP* using a scale factor of 3 and the origin as the center.

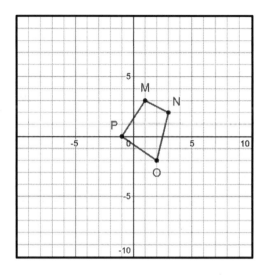

Solution:

(A)
$M(1,3) \to M'(3,9)$
$N(3,2) \to N'(9,6)$
$O(2,-2) \to O'(6,-6)$
$P(-1,0) \to P'(-3,0)$

(B)

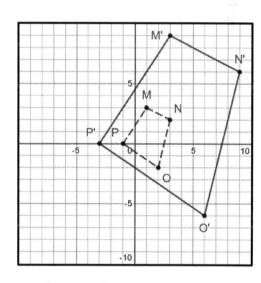

Explanation of steps:
(A) Find the image of each vertex by multiplying the coordinates of the vertex by the scale factor. *[Since the scale factor is 3, $(x,y) \to (3x, 3y)$]*
(B) Plot and label the vertices of the image and draw the sides.

Dilations
5.2 Dilations of Polygons

PRACTICE PROBLEMS

1. Triangle $A'B'C'$ is the image of $\triangle ABC$ after a dilation of 2. Which statement is true?

   (1) $AB = A'B'$

   (2) $BC = 2(B'C')$

   (3) $m\angle B = m\angle B'$

   (4) $m\angle A = \frac{1}{2}(m\angle A')$

2. Triangle ABC has vertices $A(6,6)$, $B(9,0)$, and $C(3,-3)$. State the coordinates of $\triangle A'B'C'$, the image of $\triangle ABC$ after a dilation of $D_{(0,0),\frac{1}{3}}$.

3. Graph the dilation of $\triangle ABC$ under $D_{(0,0),2}$.

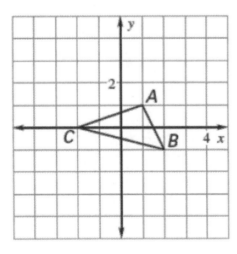

4. Graph the dilation, centered at the origin, of the quadrilateral below using a scale factor of $\frac{1}{2}$.

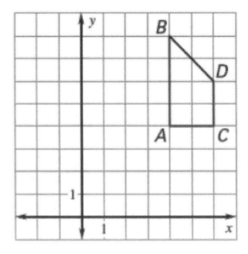

78

5. Graph △ABC with $A(-5,5)$, $B(-5,10)$, and $C(10,0)$. Then graph its dilation with a scale factor of $\frac{3}{5}$ centered at the origin.

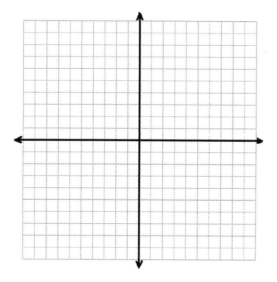

6. Triangle ABC has coordinates $A(-2,1)$, $B(3,1)$, and $C(0,-3)$. On the set of axes below, graph and label △A'B'C', the image of △ABC after a dilation of 2 centered at the origin.

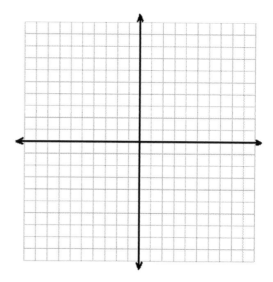

7. On the grid below, graph △ABC with coordinates $A(-1,2)$, $B(0,6)$, and $C(5,4)$. Then graph △A'B'C', the image of △ABC after a dilation of 2 centered at the origin.

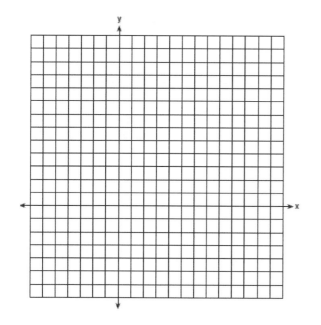

8. On the grid below, graph and label quadrilateral ABCD with coordinates $A(-1,3)$, $B(2,0)$, $C(2,-1)$, and $D(-3,-1)$. Graph, label, and state the coordinates of A'B'C'D', the image of ABCD under a dilation of 2 centered at the origin.

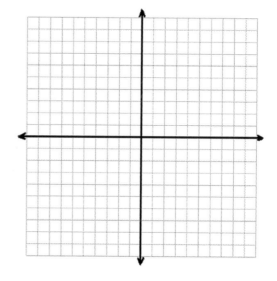

9. On the grid below, △ABC is graphed with vertices $A(-1,0)$, $B(0,4)$, and $C(2,0)$. Graph the image of △ABC after a dilation of 2 centered on point B.

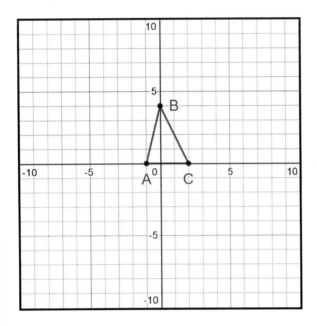

10. On the grid below, △ABC is graphed with vertices $A(0,2)$, $B(2,4)$, and $C(6,0)$. Graph the image of △ABC after a dilation of $\frac{1}{2}$ centered on point $P(2,-4)$.

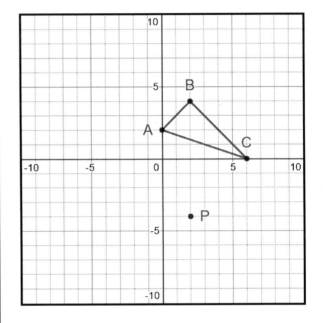

11. Given: Quadrilateral ABCD with vertices $A(-2,2)$, $B(8,-4)$, $C(6,-10)$, and $D(-4,-4)$.

(a) State the coordinates of $A'B'C'D'$, the image of quadrilateral ABCD under a dilation of factor $\frac{1}{2}$ centered at the origin.

(b) Prove that $A'B'C'D'$ is a parallelogram.

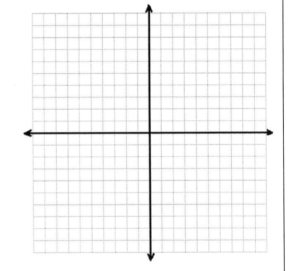

5.3 Dilations of Lines

MODEL PROBLEM

The line $y - 2x = 6$ is dilated by a scale factor of $\frac{1}{3}$ and centered at the origin. Write the equation of the image of the line after the dilation.

Solution:
(A) $y - 2x = 6 \rightarrow y = 2x + 6$
(B) Image is $y = 2x + 2$

Explanation of steps:
(A) Transform the original linear equation into slope-intercept form.
(B) The dilation of a line, with an external center point, results in a parallel line, so the slope of the image [2] remains the same. Since the center of dilation is the origin, we can find the y-intercept of the image by multiplying the original y-intercept by the scale factor [$6 \times \frac{1}{3} = 2$].

PRACTICE PROBLEMS

1. The line $y = 3x - 4$ is dilated by a scale factor of 5 and centered at the origin. What is the equation of its image?	2. The line $2x + 3y = 4$ is dilated by a scale factor of 3 and centered at the origin. What is the equation of its image?
3. The line $y = 3x - 4$ is dilated by a scale factor of 2 with a center at $C(1, -1)$. What is the equation of its image?	4. The line $y = 2x + 2$ is dilated by a scale factor of 2 with a center at $C(1, 0)$. What is the equation of its image?

Chapter 6. Transformation Proofs

6.1 Properties of Transformations

Model Problem

In $\triangle KLM$, $m\angle K = 36$ and $KM = 5$. The transformation D_2 is performed on $\triangle KLM$ to form $\triangle K'L'M'$. Find $m\angle K'$ and the length of $\overline{K'M'}$.

Solution:
(A) (B)
$m\angle K' = 36$ and $K'M' = 10$

Explanation:
(A) Dilations preserve angle measures.
[$\angle K'$ has the same measure as $\angle K$.]
(B) Dilations do not preserve distance. In fact, each side or a polygon will be enlarged or reduced based on the scale factor.
[A scale factor of 2 doubles the length of $\overline{KM}$.]

Practice Problems

1. Identify each transformation as a translation, reflection, rotation, or dilation.

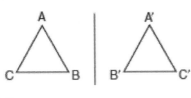

(a) _____

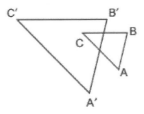

(b) _____

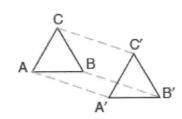

(c) _____

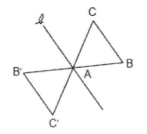

(d) _____

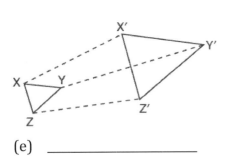

(e) _____

2. Under which transformation is △X'Y'Z' the image of △XYZ?

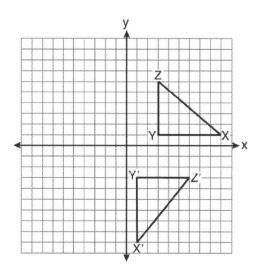

(1) dilation (3) rotation
(2) reflection (4) translation

3. Which expression best describes the transformation shown below?

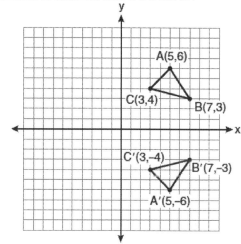

(1) same orientation; reflection
(2) opposite orientation; reflection
(3) same orientation; translation
(4) opposite orientation; translation

4. Which transformation will move $\overline{AB}$ onto $\overline{DE}$ such that point D is the image of point A and point E is the image of point B?

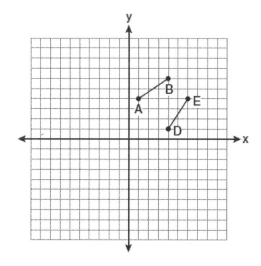

(1) $T_{3,-3}$ (3) $R_{(0,0),90°}$
(2) $D_{(0,0),\frac{1}{2}}$ (4) $r_{y=x}$

5. Under which transformation is △A'B'C' the image of △ABC?

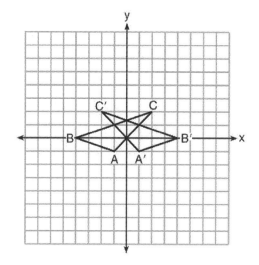

(1) $D_{(0,0),2}$ (3) r_{y-axis}
(2) r_{x-axis} (4) $(x,y) \to (x-2, y)$

6. When hexagon *ABCDEF* is reflected over line *m*, the image is hexagon *A'B'C'D'E'F'*. Under this transformation, which property is *not* preserved?

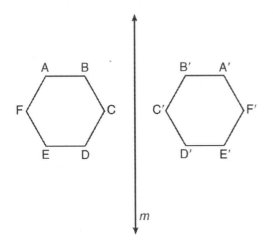

(1) area
(2) distance
(3) orientation
(4) angle measure

7. Triangle *JTM* is shown below. Which transformation would result in an image that is that is similar to, but *not* congruent to, △*JTM*?

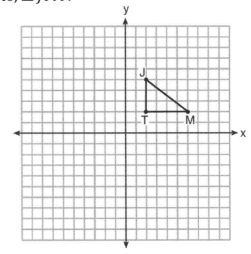

(1) $r_{y=x}$
(2) $R_{(0,0),90°}$
(3) $T_{0,-3}$
(4) $D_{(0,0),2}$

8. The perimeter of △*A'B'C'*, the image of △*ABC*, is twice as large as the perimeter of △*ABC*. What type of transformation has taken place?

(1) dilation
(2) translation
(3) rotation
(4) reflection

9. Quadrilateral *A'B'C'D'* is the image of quadrilateral *ABCD*. For which transformation would the area of *A'B'C'D'* not be equal to the area of *ABCD*?

(1) rotation of 90° about the origin
(2) reflection over the *y*-axis
(3) dilation by a scale factor of $\frac{1}{2}$
(4) translation of $(x,y) \to (x+4, y-1)$

6.2 Sequences of Transformations

Model Problem

Reflect the triangle over the line $x = -1$ and then rotate it 90° with the center at the origin.

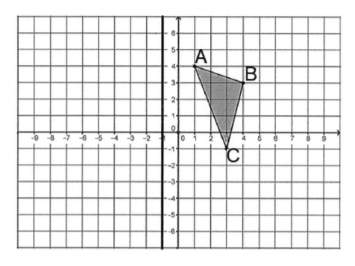

Solution:

(A)
(B)

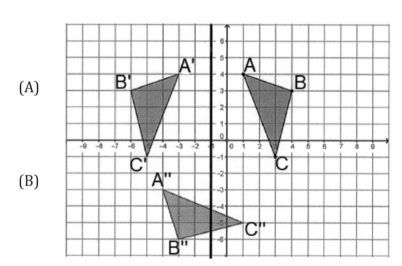

Explanation of steps:
(A) Determine the vertices after the reflection by using the line of reflection as the perpendicular bisector of $\overline{AA'}$, $\overline{BB'}$, and $\overline{CC'}$.
(B) For the rotation, use the rule $R_{(0,0), 90°} : (x, y) \rightarrow (-y, x)$.
[For example, $A'(-3,4) \rightarrow A''(-4,-3)$.]

Practice Problems

1. In the diagram to the right, △ A'B'C' is a transformation of △ ABC, and △ A''B''C'' is a transformation of △ A'B'C'.

   The composite transformation of △ ABC to △ A''B''C'' is an example of a

   (1) reflection followed by a rotation
   (2) reflection followed by a translation
   (3) translation followed by a rotation
   (4) translation followed by a reflection

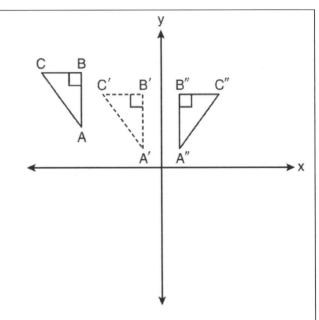

2. Describe the sequence of transformations applied in the graph below.

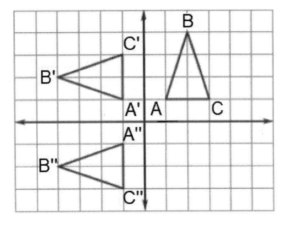

3. Describe the sequence of transformations applied in the graph below.

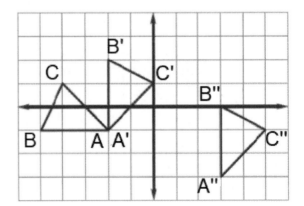

4. Describe the sequence of transformations applied in the graph below.

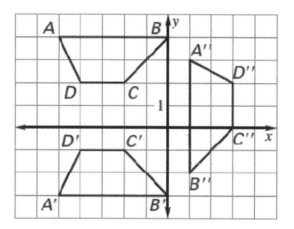

5. Graph the image of $\triangle GLQ$ after r_{y-axis} followed by $T_{4,-4}$.

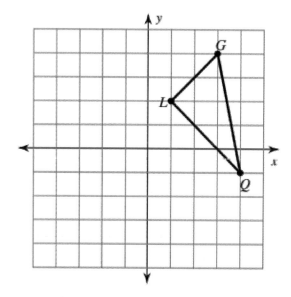

6. Graph the image of $\triangle LUX$ after $r_{x=1}$ followed by $D_{(0,0),\frac{2}{5}}$.

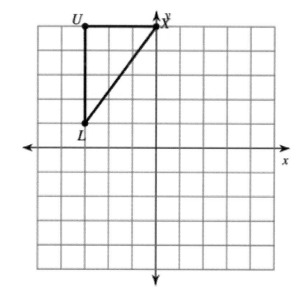

7. Graph the image of $MPDZ$ after $T_{-3,0}$ followed by $R_{(0,0),180°}$.

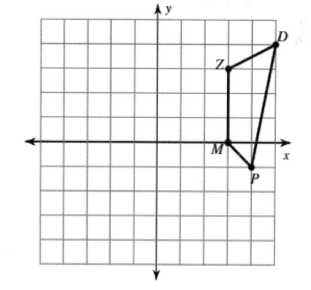

8. The point (3, −2) is rotated 90° about the origin and then dilated by a scale factor of 4. What are the coordinates of the resulting image?

   (1) (−12, 8) (3) (8, 12)
   (2) (12, −8) (4) (−8, −12)

9. The coordinates of △ABC, shown on the graph below, are A(2,5), B(5,7), and C(4,1).

   a) Graph and label △A'B'C', the image of △ABC after a reflection over the y-axis.
   b) Graph and label △A''B''C'', the image of △A'B'C' after a reflection over the x-axis.
   c) State a single transformation that will map △ABC onto △A''B''C''.

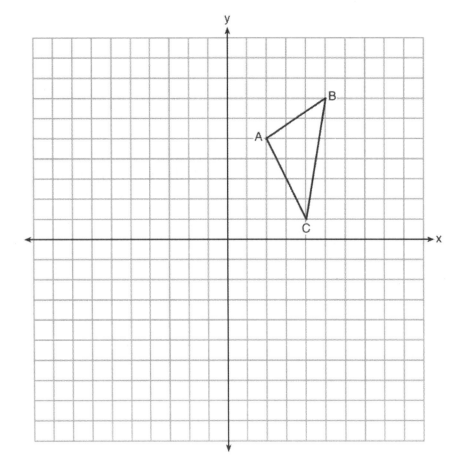

10. Given △ ABC with points A(4,3), B(4,−2), and C(2,3).

   a) On the grid below, sketch △ ABC.
   b) On the same set of axes, graph and state the coordinates of △ A'B'C', the image of △ ABC after a reflection in the line $y = x$.
   c) On the same set of axes, graph and state the coordinates of △ A''B''C'', the image of △ A'B'C' after the translation $T_{-4,3}$.

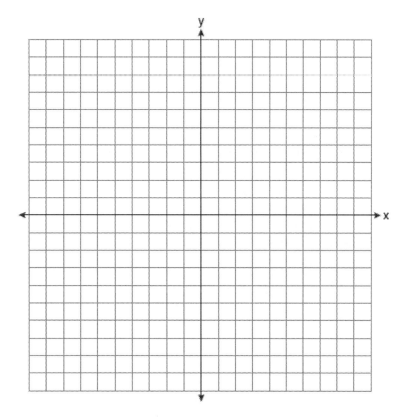

11. The coordinates of the endpoints of $\overline{AB}$ are $A(2,6)$ and $B(4,2)$. Is the image $\overline{A''B''}$ the same if it is reflected in the x-axis, then dilated by $\frac{1}{2}$ centered on the origin as the image is if it is dilated by $\frac{1}{2}$ centered on the origin, then reflected in the x-axis? Justify your answer.

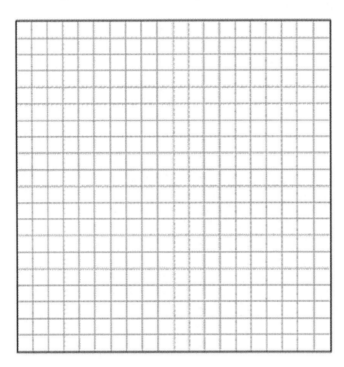

6.3 Transformations and Congruence

MODEL PROBLEM

In the diagram below, △ ABC ≅ △ DEF. Describe a sequence of rigid motions that will map △ ABC onto △ DEF.

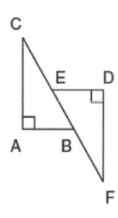

Solution:
Translate △ ABC so that B coincides with E.
Then rotate △ ABC 180° around B.

Explanation:
See diagram at right.

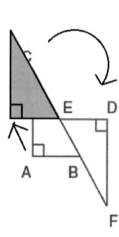

PRACTICE PROBLEMS

1. Given △ ABC ≅ △ DEF and $\overline{ADBE}$. Describe a rigid motion that maps △ ABC onto △ DEF.

   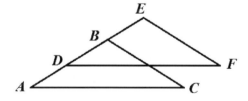

2. Given △ ABC ≅ △ ADC. Describe a rigid motion that maps △ ABC onto △ ADC.

   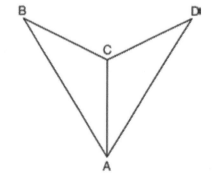

3. Given $\overline{HK}$ and $\overline{IL}$, $\triangle HIJ \cong \triangle KLJ$. Describe a rigid motion that maps $\triangle HIJ$ onto $\triangle KLJ$.

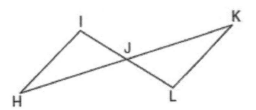

4. Given $\triangle HIK \cong \triangle KJH$. Describe a rigid motion that maps $\triangle HIK$ onto $\triangle KJH$.

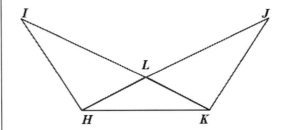

5. Given $\triangle PRQ \cong \triangle ABC$. Describe a sequence of rigid motions that maps $\triangle PRQ$ to $\triangle ABC$. Is it possible to show the congruence using only translations and rotations?

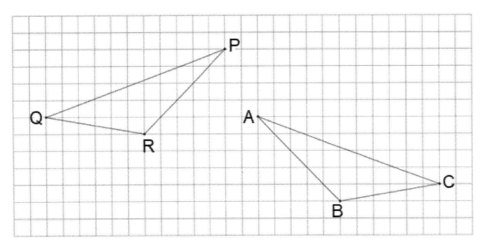

6.4 Transformations and Similarity

MODEL PROBLEM

Prove $\triangle LMN \sim \triangle STR$ by giving a sequence of transformations that maps $\triangle LMN$ onto $\triangle STR$.

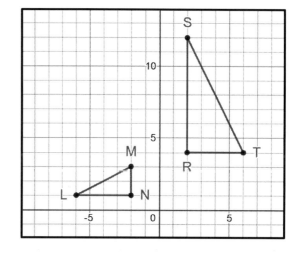

Solution:
(A)   (B)
Rotation $R_{(0,0),270°}$ followed by dilation $D_{(0,0),2}$.

Explanation of steps:
(A) To align $\triangle LMN$ to the same direction as $\triangle STR$, we need to rotate it.
[We can either rotate it 90° clockwise or 270° counterclockwise to achieve the same result. Since $R_{(0,0),270°} : (x,y) \to (y,-x)$, this will map $N(-2,1) \to N'(1,2)$.]

(B) Since $\triangle STR$ is larger than $\triangle LMN$, a dilation is needed. *[A dilation of scale factor 2 centered on the origin would map $N'(1,2)$ to $R(2,4)$. This will also map L' onto S and M' onto T.]*

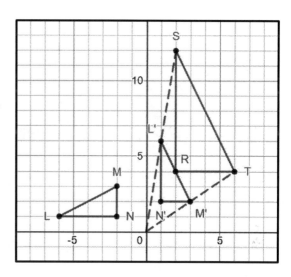

Practice Problems

1. Prove △ABC ~ △A'B'C' by giving a sequence of transformations that maps △ABC onto △A'B'C'.

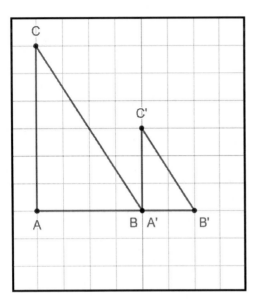

2. Prove △ABC ~ △A'B'C' by giving a sequence of transformations that maps △ABC onto △A'B'C'.

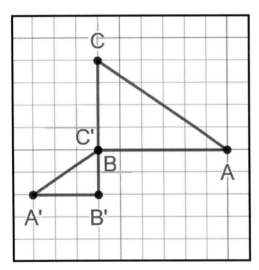

Chapter 7. Circles in the Coordinate Plane

7.1 Equation of a Circle

Model Problem

What is the center and radius of the circle whose equation is $x^2 - 2x + y^2 + 6y - 6 = 0$?

Solution:

(A) $x^2 - 2x + y^2 + 6y - 6 = 0$

(B) $x^2 - 2x + y^2 + 6y = 6$

(C) $\left(\dfrac{h}{2}\right)^2 = \left(\dfrac{-2}{2}\right)^2 = 1$

$(x^2 - 2x + 1) + y^2 + 6y = 6 + 1$

(D) $(x-1)^2 + y^2 + 6y = 7$

(E) $\left(\dfrac{k}{2}\right)^2 = \left(\dfrac{6}{2}\right)^2 = 9$

$(x-1)^2 + (y^2 + 6y + 9) = 7 + 9$

(F) $(x-1)^2 + (y+3)^2 = 16$

(G) Center is $(1, -3)$ and radius is 4.

Explanation of steps:

(A) Write the equation in the form, $x^2 + hx + y^2 + ky + c = 0$. *[The equation was already in this form.]*

(B) Add the opposite of c to both sides. *[Add 6 to both sides.]*

(C) Add $\left(\dfrac{h}{2}\right)^2$ to both sides. *[Write this value, 1, next to the x terms to create a trinomial.]*

(D) Factor the trinomial involving x into a binomial squared.

(E) Add $\left(\dfrac{k}{2}\right)^2$ to both sides. *[Write this value, 9, next to the y terms to create a trinomial.]*

(F) Factor the trinomial involving y into a binomial squared.

(G) Now that the equation is in center-radius form, $(x-a)^2 + (y-b)^2 = r^2$, state the center (a, b) and the radius r. *[$r^2 = 16$, so $r = \sqrt{16} = 4$.]*

Practice Problems

1. A circle has the equation $x^2 + y^2 = 10$. a) What are the coordinates of its center and the length of its radius? b) Is the point $(3, -1)$ on the circle?	2. A circle has the equation $(x-2)^2 + (y+3)^2 = 36$. What are the coordinates of its center and the length of its radius?

Circles in the Coordinate Plane — 7.1 Equation of a Circle

3. A circle has the equation $(x-1)^2 + (y+3)^2 = 9$.

   What are the coordinates of its center and the length of its radius?

4. A circle has the equation $x^2 + (y-7)^2 = 32$.

   What are the coordinates of its center and the length of its radius in simplest radical form?

5. What is the equation of a circle whose center is the origin and that passes through the point $(-4, 0)$?

6. A circle whose center is $(-3, 4)$ passes through the origin. What is the equation of the circle?

7. What is the equation of the circle graphed below, with the center and a point on the circle given?

   Points shown: $(1, -2)$ and $(4, -2)$

8. The coordinates of the endpoints of the diameter of a circle are $(2, 0)$ and $(2, -8)$. What is the equation of the circle?

Circles in the Coordinate Plane — 7.1 Equation of a Circle

9. Circle O has the equation $(x-3)^2 + (y+2)^2 = 36$.

   Circle O' is the image of circle O after a dilation of D_3. What are the coordinates of the center of circle O'?

10. Circle O has the equation $(x+4)^2 + (y-2)^2 = 81$.

    Circle O' is the image of circle O after a dilation of D_2. What are the coordinates of the center and the length of the radius of circle O'?

11. A circle has the equation $x^2 + y^2 + 4x = 5$.

    What are the coordinates of its center and the length of its radius?

12. A circle has the equation $x^2 + y^2 + 6x - 4y = 12$.

    What are the coordinates of its center and the length of its radius?

13. A circle has the equation
    $x^2 + y^2 - 16x + 6y + 53 = 0$.

    What are the coordinates of its center and the length of its radius in simplest radical form?

14. A circle has the equation
    $x^2 - 2x = -y^2 + 10y + 1$.

    What are the coordinates of its center and the length of its radius in simplest radical form?

15. In the diagram below, circle O is drawn such that the y-axis is tangent to the circle at A and the x-axis is tangent to the circle at B. Point $(6,3)$ lies on $\widehat{AB}$. Find r, the radius of the circle. *[Hint: it can be determined from the diagram that r must be greater than 6.]*

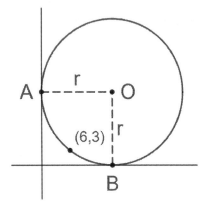

7.2 Graph Circles

Model Problem

Graph the circle whose equation is $(x + 2)^2 + (y - 1)^2 = 25$.

Solution:

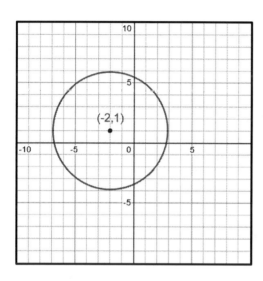

Explanation of steps:
- (A) Write the equation in center-radius form. *[The equation is already in this form.]*
- (B) Graph the center point at (a, b). *[The center is $(-2, 1)$]*
- (C) Open the compass to r units wide. *[$r = \sqrt{25} = 5$. If the compass point is placed at the center $(-2, 1)$ and the compass pencil is stretched to $(-2, 6)$, this would be 5 units wide.]*
- (D) Place the compass point on the center and draw a circle with radius of r.

Practice Problems

1. On the grid below, draw a circle whose equation is $x^2 + y^2 = 25$.

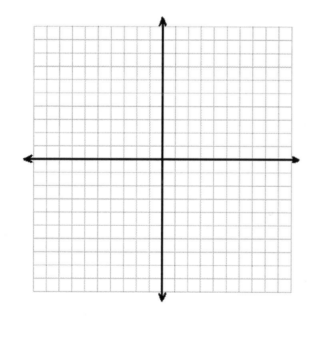

2. On the grid below, draw a circle whose equation is $(x - 5)^2 + (y + 3)^2 = 16$.

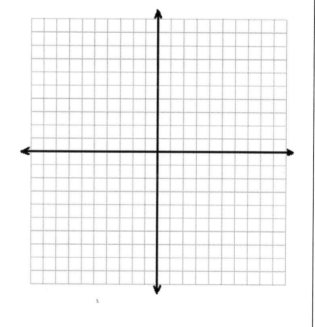

3. On the grid below, draw the circle whose equation is $(x-2)^2 + y^2 = 9$.

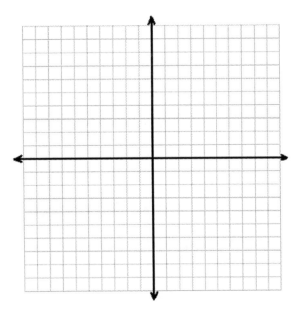

4. On the grid below, draw the circle whose equation is $x^2 + (y+1)^2 = 36$.

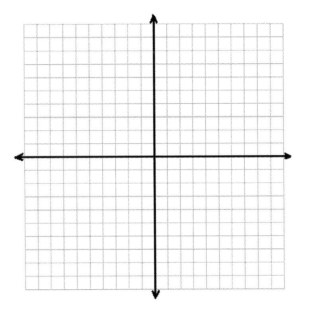

5. On the grid below, draw the circle whose equation is $x^2 + 4x + y^2 - 4y = 41$.

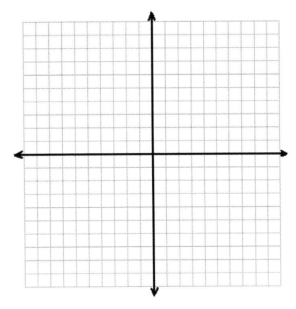

Chapter 8. Foundations of Euclidean Geometry

8.1 Postulates, Theorems and Proofs

Model Problem

Given m∠PTQ = m∠STR, prove m∠PTR = m∠STQ.

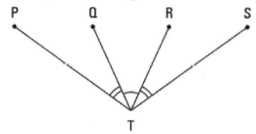

Solution:

Statements	Reasons
m∠PTQ = m∠STR	Given
m∠QTR = m∠QTR	Reflexive Property
m∠PTQ + m∠QTR = m∠STR + m∠QTR	Addition Property
m∠PTR = m∠PTQ + m∠QTR and m∠STQ = m∠STR + m∠QTR	Angle Addition Postulate
m∠PTR = m∠STQ	Substitution

Explanation of steps:
Start with the Given statement and follow a logical sequence of deductions until the statement we need to prove is the final statement.
[When looking at the two angles that we need to prove equal in measure, ∠PTR and ∠STQ, we notice that they share a common angle, ∠QTR. Namely, we get m∠PTR by adding m∠PTQ + m∠QTR, and we get m∠STQ by adding m∠STR + m∠QTR, by the Angle Addition Postulate. Working backwards, if we can show these angle sums are equal, then m∠PTR = m∠STQ by Substitution. We do know the two sums are equal because we are adding equals to equals (the Addition Property): m∠PTQ = m∠STR by the Given statement, and m∠QTR is equal to itself by the Reflexive Property.]

Practice Problems

1. When writing a geometric proof, which angle relationship could be used alone to justify that two angles are congruent?

   (1) supplementary angles
   (2) linear pair of angles
   (3) adjacent angles
   (4) vertical angles

2. In △ AED with $\overline{ABCD}$ shown in the diagram below, $\overline{EB}$ and $\overline{EC}$ are drawn.

   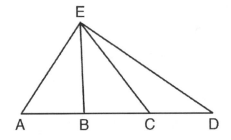

   If $\overline{AB} \cong \overline{CD}$, which statement could always be proven?

   (1) $\overline{AC} \cong \overline{DB}$ (3) $\overline{AB} \cong \overline{BC}$
   (2) $\overline{AE} \cong \overline{ED}$ (4) $\overline{EC} \cong \overline{EA}$

3. In the diagram below of $\overline{ABCD}$, $\overline{AC} \cong \overline{BD}$.

   Using this information, it could be proven that

   (1) $BC = AB$ (3) $AD - BC = CD$
   (2) $AB = CD$ (4) $AB + CD = AD$

4. In the diagram of $\overline{WXYZ}$ below, $\overline{WY} \cong \overline{XZ}$.

Which reasons can be used to prove $\overline{WX} \cong \overline{YZ}$?

   (1)   reflexive property and addition property
   (2)   reflexive property and subtraction property
   (3)   transitive property and addition property
   (4)   transitive property and subtraction property

5. Given: ∠1 and ∠2 are complementary
            ∠2 and ∠3 are complementary
   Prove: m∠1 = m∠3

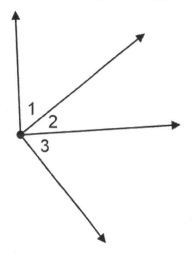

Statements	Reasons

6. Given: $\overline{PCEG}$
   $\overline{PC} \cong \overline{GE}$
   Prove: $\overline{PE} \cong \overline{GC}$

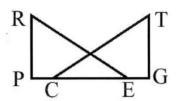

Statements	Reasons

8.2 Parallel Lines and Transversals

Model Problem

Lines *s* and *t* are parallel. Find the measures of the angles labelled 1, 2, 3, and 4.

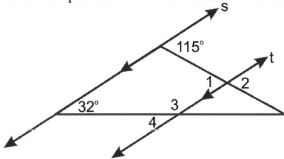

Solution:
(A) m∠1 = 115°
(B) m∠2 = 115°
(C) m∠4 = 32°
(D) m∠3 = 148°

Explanation of steps:
(A) ∠1 and the angle marked as 115° are alternate interior angles, which are congruent.
(B) ∠1 and ∠2 are vertical angles, which are congruent.
(C) ∠4 and the angle marked as 32° are alternate interior angles, which are congruent.
(D) ∠3 and ∠4 are a linear pair, which are supplementary (add to 180°).

Practice Problems

1. In the diagram of lines ℓ, m. and n, which is a pair of corresponding angles?

   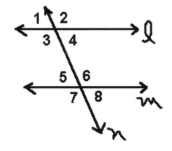

   (1) ∠1 and ∠8   (3) ∠4 and ∠8
   (2) ∠3 and ∠6   (4) ∠4 and ∠6

2. In the diagram of lines ℓ, m. and n, which is a pair of alternate exterior angles?

   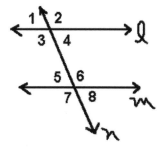

   (1) ∠1 and ∠8   (3) ∠4 and ∠8
   (2) ∠3 and ∠6   (4) ∠4 and ∠6

3. Given: $r \parallel s$
   State the relationship between each pair of angles as shown in the diagram, and state whether the angles are congruent or supplementary.

   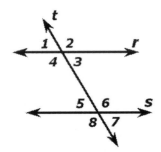

   a) ∠1 and ∠2
   b) ∠2 and ∠4
   c) ∠2 and ∠6
   d) ∠4 and ∠6
   e) ∠2 and ∠8

4. If two parallel lines are intersected by a transversal, which of the following pairs of angles is always supplementary?

   (1) corresponding angles
   (2) consecutive interior angles
   (3) alternate interior angles
   (4) alternate exterior angles

5. A transversal intersects two lines. Which condition would always make the two lines parallel?

   (1) Vertical angles are congruent.
   (2) Alternate interior angles are congruent.
   (3) Corresponding angles are supplementary.
   (4) Same-side interior angles are complementary.

6. $\overleftrightarrow{EF}$ intersects $\overleftrightarrow{AB}$ and $\overleftrightarrow{CD}$, as shown in the diagram below.

   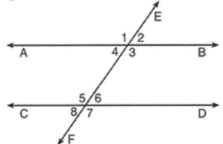

   Which statement could always be used to prove $\overleftrightarrow{AB} \parallel \overleftrightarrow{CD}$?

   (1) ∠2 ≅ ∠4
   (2) ∠7 ≅ ∠8
   (3) ∠3 and ∠6 are supplementary
   (4) ∠1 and ∠5 are supplementary

7. Based on the diagram below, which statement must be true?

   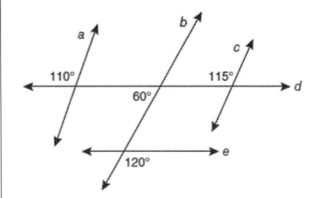

   (1) $a \parallel b$ (3) $b \parallel c$
   (2) $a \parallel c$ (4) $d \parallel e$

8. In the diagram, two parallel lines are intersected by a transversal and ∠1 and ∠2 are labelled. If m∠1 = $(x + 20)°$ and m∠2 = $(2x - 10)°$, find x.

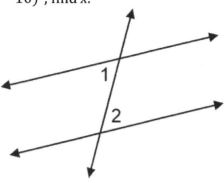

9. Parallel lines $\overleftrightarrow{AB}$ and $\overleftrightarrow{CD}$ are intersected by $\overleftrightarrow{EF}$ at points X and Y, and m∠FYD = 123°. Find m∠AXY.

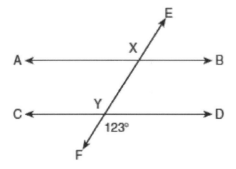

10. In the diagram below, lines n and m are cut by transversals p and q. What value of x would make lines n and m parallel?

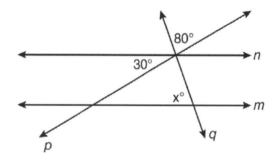

11. Lines m and n are parallel. Find the measures of the angles labelled a, b, c, and d.

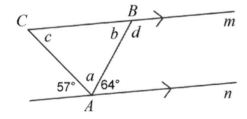

12. Given $p \parallel q$, solve for x and y.

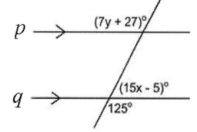

13. Given: ∠1 and ∠3 are supplementary
    Prove: $m \parallel n$

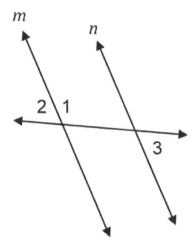

Statements	Reasons

Chapter 9. Triangles and Congruence

9.1 Angles of Triangles

Model Problem

In the diagram to the right, $\overleftrightarrow{ABCD}$ is a straight line, and $\angle E$ in triangle BEC is a right angle.

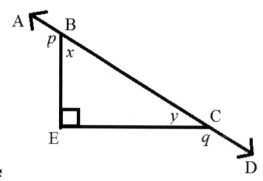

How many degrees are in $p + q$?

Solution:
(A) $p = 90 + y$
$q = 90 + x$
(B) $x + y = 90$
(C) Therefore,
$p + q = 90 + y + 90 + x$
$= 180 + x + y$
$= 180 + 90$
$= 270$
So, $p + q = 270°$

Explanation of steps:
(A) The exterior angle of a triangle equals the sum of the two remote interior angles. *[p and q are exterior angles]*
(B) In a right triangle, the two acute angles are complementary.
(C) Find the sum *[by substituting for p, substituting for q, and then substituting for x + y]*.

Practice Problems

1. What is the measure of angle C?

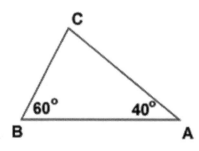

2. In the diagram of $\triangle KLM$ below, $m\angle L = 70$, $m\angle M = 50$, and $\overline{MK}$ is extended through N. Find $m\angle LKN$.

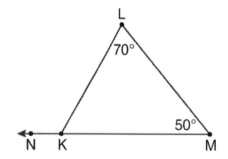

3. In the diagram below, $\overleftrightarrow{RCBT}$ and $\triangle ABC$ are shown with m∠A = 60 and m∠ABT = 125. What is m∠ACR?

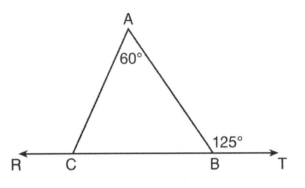

4. What is the measure of an exterior angle of an equilateral triangle?

5. Solve for x.

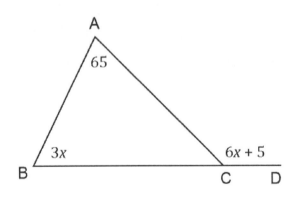

6. The measure of the largest angle of a triangle is 5 times the smallest angle. The third angle is 12 degrees larger than the smallest angle. Is the triangle acute, right, or obtuse?

7. In the diagram of △JEA below, m∠JEA = 90° and m∠EAJ = 48°. $\overline{MS}$ connects points M and S on the triangle, such that m∠EMS = 59°. What is m∠JSM?

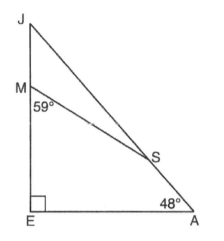

8. In the accompanying diagram, $\overleftrightarrow{AB} \parallel \overleftrightarrow{CD}$. From point E on $\overleftrightarrow{AB}$, transversals $\overrightarrow{EF}$ and $\overrightarrow{EG}$ are drawn, intersecting $\overleftrightarrow{CD}$ at H and I, respectively. If m∠CHF = 20 and m∠DIG = 60, what is m∠HEI?

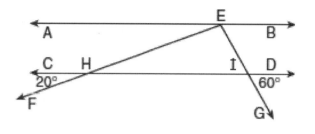

9. In the diagram below, ℓ ∥ m and $\overline{QR} \perp \overline{ST}$ at R. If m∠1 = 63, find m∠2.

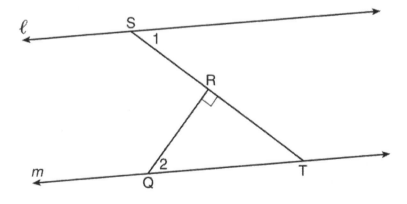

10. The diagram below shows △ ABD, with $\overrightarrow{ABC}$, $\overline{BE} \perp \overline{AD}$, and ∠EBD ≅ ∠CBD. If m∠ABE = 52, what is m∠D?

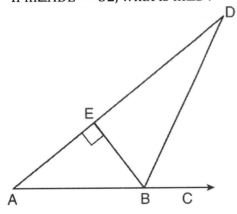

11. In the diagram below, m∠B = 90°, m∠A = 65°, m∠D = 50°, and m∠DCE = 80°. Find m∠x.

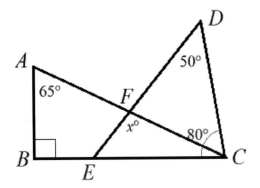

12. In the diagram below, $a \parallel b$. Find m∠x.

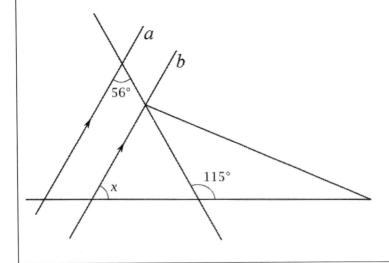

13. In the diagram below of quadrilateral *ABCD* with diagonal $\overline{BD}$, m∠A = 93, m∠ADB = 43, m∠C = 3x + 5, m∠BDC = x + 19, and m∠DBC = 2x + 6.
Determine if $\overline{AB}$ is parallel to $\overline{DC}$. Explain your reasoning.

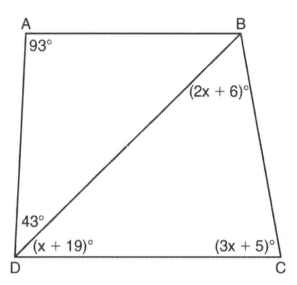

Triangles and Congruence 9.2 Triangle Inequality Theorem

9.2 Triangle Inequality Theorem

MODEL PROBLEM

The measures of two sides of a triangle are 10 and 15. Use a compound inequality to express the range of the measure of the third side, c.

Solution:
$5 < c < 25$

Explanation of steps:
(A) The larger side *[15]* must be less than the sum of the other two sides, so the third side must be larger than the difference *[$c > 15 - 10$ or $c > 5$]*.
(B) The third side must also be less than the sum of the other two sides *[so, $c < 10 + 15$ or $c < 25$]*.
(C) Combine the two inequalities into a compound inequality *[$c > 5$ and $c < 25$ means $5 < c < 25$]*.

PRACTICE PROBLEMS

1. Which of these lengths could be the sides of a triangle?

   (1) 15 cm, 7 cm, 23 cm
   (2) 5 cm, 9 cm, 13 cm
   (3) 8 cm, 5 cm, 13 cm
   (4) 6 cm, 15 cm, 23 cm

2. The lengths of two sides of a triangle are 7 and 11. Which inequality represents all possible values for x, the length of the third side of the triangle?

   (1) $4 \leq x \leq 18$
   (2) $4 < x \leq 18$
   (3) $4 \leq x < 18$
   (4) $4 < x < 18$

3. The lengths of two sides of a triangle are 11 and 15.
   Which inequality gives the range of possible lengths of the third side, x.

   (1) $4 < x < 26$
   (2) $11 < x < 26$
   (3) $0 < x < 10$
   (4) $1 < x < 17$

4. Rae wants to build a triangular pen for her pet iguana. She has three boards already cut that measure 7 feet, 8 feet, and 16 feet in length.

   Explain why Rae cannot construct a triangular pen with sides of 7 feet, 8 feet, and 16 feet.

5. In the diagram of △ABC, D is a point on $\overline{AB}$, AC = 7, AD = 6, and BC = 18.

   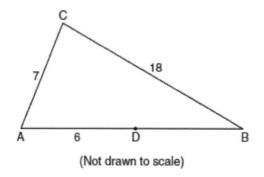
   (Not drawn to scale)

   The length of $\overline{DB}$ could be
   (1) 5 (3) 19
   (2) 12 (4) 25

6. The triangle in the diagram has a perimeter of 34 yards. Find the length, in yards, of each side of the figure.

   Could these measures actually represent the measures of the sides of a triangle? Explain your answer.

   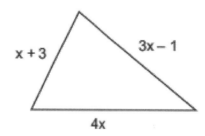

7. In △ABC, AB = 5, BC = 8, and AC = 10. Arrange the angles of the triangle from largest to smallest.

8. In △DEF, m∠D = 125°, m∠E = 25°, and m∠F = 30°. Arrange the sides of the triangle from longest to shortest.

9. Given $\overline{ABC}$, m∠A = 66°, m∠CDB = 18°, and m∠C = 24°. What are the longest and shortest sides of △ABD?

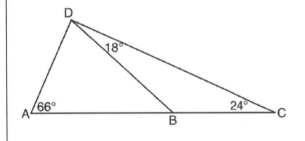

10. Given m∠A = 57°, m∠ABD = 87°, m∠CBD = 59°, and m∠C = 28°. Which segment in the diagram is the shortest?

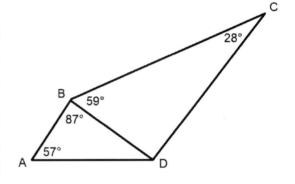

9.3 Segments in Triangles

Model Problem

$\overline{BD}$ is a median of $\triangle ABC$. If $AD = 10x - 7$, $DC = 5x + 3$, and $m\angle BDC = (15x + 42)°$, find $m\angle BDC$.

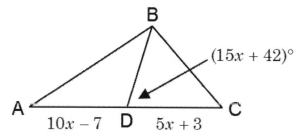

Solution:
(A) $10x - 7 = 5x + 3$
$5x = 10$
$x = 2$
(B) $15(2) + 42 = 72°$

Explanation of steps:
(A) A median is drawn to the midpoint of a side, which divides the side into two congruent parts. *[D is the midpoint of $\overline{AC}$, so $AD = DC$.]*
(B) Substitute the value of x into the expression of the measure we need to find. *[Substitute $x = 2$ into the expression for the measure of $\angle BDC$.]*

Practice Problems

1. In the diagram of $\triangle ABC$, name:

    a) a median

    b) an altitude

    c) an angle bisector

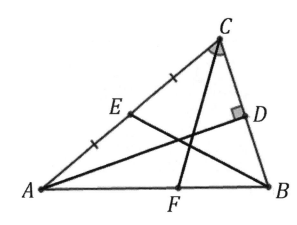

2. Given △ABC with median $\overline{BF}$, altitude $\overline{BD}$, and angle bisector $\overline{BE}$ drawn to side $\overline{AC}$, which conclusion is valid?

    (1) ∠FAB ≅ ∠ABF
    (2) ∠ABF ≅ ∠CBD
    (3) $\overline{CE} \cong \overline{EA}$
    (4) $\overline{CF} \cong \overline{FA}$

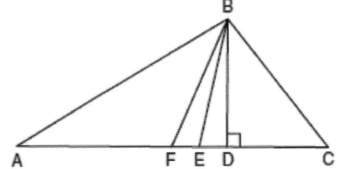

3. $\overline{YW}$ is an altitude of △XYZ. If m∠YWX = $(6x - 6)°$, find x.

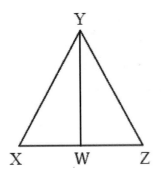

4. $\overline{YW}$ bisects ∠XYZ. If m∠XYW = $4x - 17$ and m∠ZYW = $3x - 4$, find m∠XYZ.

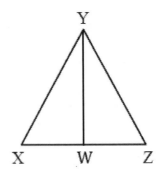

5. $\overline{SQ}$ is an altitude of $\triangle PRS$. If $m\angle QSR = 4x - 8$ and $m\angle R = 6x + 13$, find x.

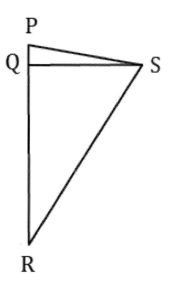

6. $\overline{AE}$ and $\overline{DB}$ are medians of $\triangle ACD$. If $AB = 6x + 10$, $BC = x^2 + 3x$, $CD = 12y + 24$, and $DE = 2y + 60$, find x and y.

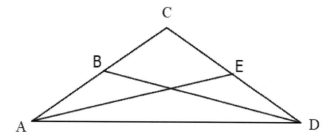

7. In △DEG below, $\overleftrightarrow{FH}$ is the perpendicular bisector of $\overline{DHG}$.

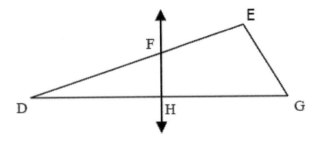

a) If $DH = 2x + 3$ and $GH = 7x - 47$, find the length of $\overline{DG}$.

b) If m∠FHG = $(y^2 + 9)°$ and m∠EFH = $(12y)°$, find m∠EFH.

9.4 Isosceles and Equilateral Triangles

Model Problem

In the diagram at right of $\triangle CAB$ and $\triangle CDB$, $\overline{AC} \cong \overline{AB}$, $\overline{DC} \cong \overline{DB}$, and $\angle CDB$ is a right angle.
Find x, the measure of $\angle ACD$.

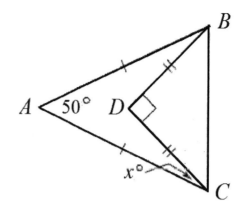

Solution:

(A) $m\angle DCB = \dfrac{180 - 90}{2} = 45°$

(B) $m\angle ACB = \dfrac{180 - 50}{2} = 65°$

(C) $x = m\angle ACB - m\angle DCB = 65 - 45 = 20°$

Explanation of steps:

(A) Use the formula $b = \dfrac{180 - v}{2}$ to calculate the base angle *[of $\triangle CDB$]*.

(B) Use the formula $b = \dfrac{180 - v}{2}$ to calculate the base angle *[of $\triangle CAB$]*.

(C) Use the Partition Postulate: subtract the part [m∠DCB] from the whole [m∠ACB].

Practice Problems

1. For an isosceles triangle whose vertex angle measures 120°, what is the measure of each base angle?	2. In $\triangle RST$, $m\angle RST = 46$ and $\overline{RS} \cong \overline{ST}$. Find $m\angle STR$.

3. In the diagram below, identify:

   a) an angle bisector

   b) an altitude

   c) a median

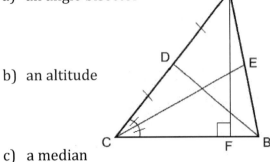

4. In all isosceles triangles, the exterior angle of a base angle must always be

   (1) a right angle
   (2) an acute angle
   (3) an obtuse angle
   (4) equal to the vertex angle

5. In the diagram of isosceles △ABC, the measure of vertex angle B is 80°. If $\overline{AC}$ extends to point D, what is m∠BCD?

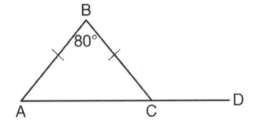

6. In the diagram below, △LMO is isosceles with LO = MO.

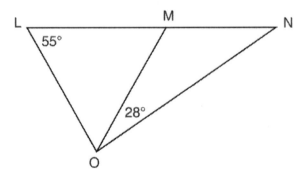

If m∠L = 55 and m∠NOM = 28, what is m∠N?

7. For the diagram below, find:

   a) m∠BCA   b) m∠DCE   c) m∠BCD

   d) m∠DEF   e) m∠BAG   f) m∠GAH

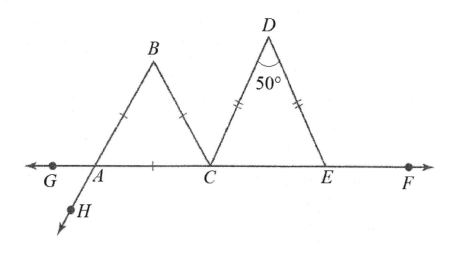

8. For the diagram of triangles PQR and QRS below, where $\overline{PRS}$ is a straight line segment, find x.

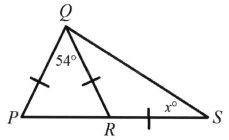

9. For the diagram of triangles DEF and EFG below, find x.

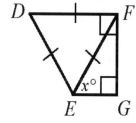

123

10. In the diagram of △ACD, B is a point on $\overline{AC}$ such that △ADB is an equilateral triangle, and △DBC is an isosceles triangle with $\overline{DB} \cong \overline{BC}$. Find m∠C.

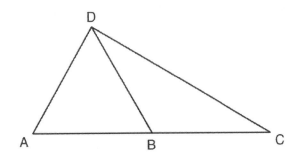

11. In the diagram below of △GJK, H is a point on $\overline{GJ}$, $\overline{HJ} \cong \overline{JK}$, m∠G = 28, and m∠GJK = 70. Determine whether △GHK is an isosceles triangle and justify your answer.

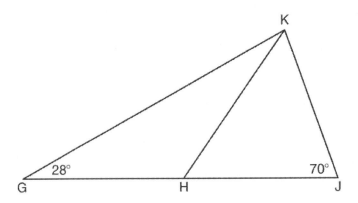

12. The diagram below shows a pennant in the shape of an isosceles triangle. Each side measures 13, the altitude is $x + 7$, and the base is $2x$. What is the length of the base?

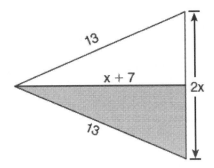

13. Given: △JKL, side $\overline{JL}$ extended through L to M, side $\overline{KL}$ extended through L to N, m∠K = 70°, m∠MLN = 55°
    Prove: △JKL is isosceles.

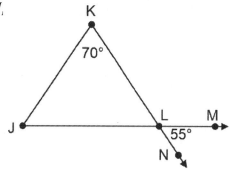

Statements	Reasons
m∠K = 70°, m∠MLN = 55°	Given

9.5 Triangle Congruence Methods

Model Problem

In the diagram below, where $\overline{LP}$ intersects $\overline{NQ}$ at M, we are given $\overline{LM} \cong \overline{PM}$ and $\overline{NM} \cong \overline{QM}$. Suppose we want to prove $\triangle LMN \cong \triangle PMQ$.
a) Name an additional pair of corresponding parts that must be congruent, and explain why.
c) State the congruence method we could then use to prove the triangles congruent.

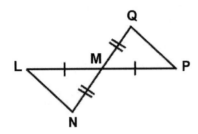

Solution:
- (A) $\angle LMN \cong \angle PMQ$ because vertical angles are congruent.
- (B) The triangles are congruent by SAS.

Explanation of steps:
- (A) If we need an additional pair of corresponding parts congruent, look for what we can deduce from the given information or from the diagram itself.
  [Intersecting lines form congruent vertical angles.]
- (B) Trace the pairs of congruent parts in the proper corresponding order to determine which congruence method to use.
  [Tracing from vertex L to M to N gives us a side (1 tick), an angle ($\angle LMN$), and another side (2 ticks). Tracing from vertex P to M to Q gives us their congruent corresponding parts: a side (1 tick), an angle ($\angle PMQ$), and a side (2 ticks). Therefore, they are congruent by SAS.]

Practice Problems

1. Given the diagram below, which statement properly uses corresponding order?

    (1) △ FET ≅ △ WRY
    (2) △ FET ≅ △ YRW
    (3) △ EFT ≅ △ YRW
    (4) △ EFT ≅ △ WRY

    By which congruence method are the two triangles congruent?

    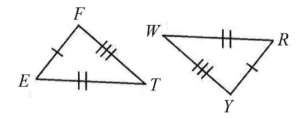

2. Given the diagram below, which statement properly uses corresponding order?

    (1) △ UIA ≅ △ OEY
    (2) △ UIA ≅ △ OYE
    (3) △ AUI ≅ △ EOY
    (4) △ IAU ≅ △ OYE

    By which congruence method are the two triangles congruent?

    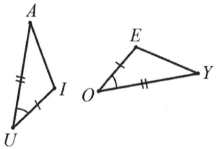

3. If △ ABC ≅ △ XYZ, which statement is always true?

    (1) ∠C ≅ ∠Y   (3) $\overline{AC} \cong \overline{YZ}$
    (2) ∠A ≅ ∠X   (4) $\overline{CB} \cong \overline{XZ}$

4. If △ JKL ≅ △ MNO, which statement is always true?

    (1) ∠KLJ ≅ ∠NMO   (3) $\overline{JL} \cong \overline{MO}$
    (2) ∠KJL ≅ ∠MON   (4) $\overline{JK} \cong \overline{ON}$

5. In the diagram of triangles *BAT* and *FLU*, ∠B ≅ ∠F and $\overline{BA} \cong \overline{FL}$.

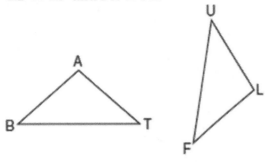

Which statement is needed to prove △ BAT ≅ △ FLU?

(1) ∠A ≅ ∠L (3) ∠A ≅ ∠U
(2) $\overline{AT} \cong \overline{LU}$ (4) $\overline{BA} \parallel \overline{FL}$

6. In the diagram of △ AGE and △ OLD, ∠GAE ≅ ∠LOD, and $\overline{AE} \cong \overline{OD}$.

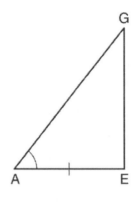 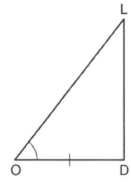

To prove that △ AGE and △ OLD are congruent by SAS, what other information is needed?

7. In the diagram of △ ABC and △ DEF, $\overline{AB} \cong \overline{DE}$, ∠A ≅ ∠D, and ∠B ≅ ∠E. Which congruence method can be used to prove △ ABC ≅ △ DEF?

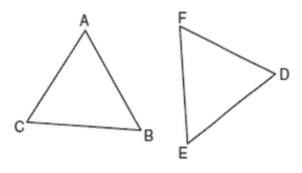

8. In the diagram of △ DAE and △ BCE, $\overline{AB}$ and $\overline{CD}$ intersect at E, such that $\overline{AE} \cong \overline{CE}$ and ∠BCE ≅ ∠DAE. Which congruence method can be used to prove △ DAE ≅ △ BCE?

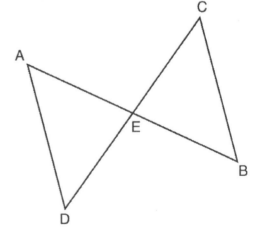

9. In the diagram, $\overline{HK}$ bisects $\overline{IL}$ and $\angle H \cong \angle K$. What is the most direct congruence method that could be used to prove $\triangle HIJ \cong \triangle KLJ$?

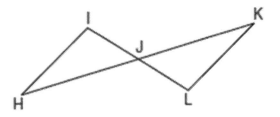

10. In the diagram, $\overline{AB} \cong \overline{CD}$ and $\angle BAC \cong \angle DCA$. What is the most direct congruence method that could be used to prove $\triangle ABC \cong \triangle CDA$?

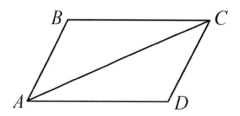

11. As shown in the diagram below, $\overline{AC}$ bisects $\angle BAD$ and $\angle B \cong \angle D$. What is the most direct congruence method that could be used to prove $\triangle ABC$ is congruent to $\triangle ADC$?

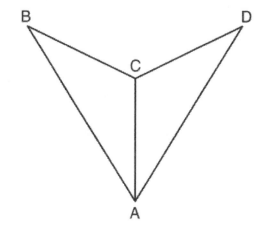

12. In the diagram of quadrilateral ABCD, $\overline{AB} \parallel \overline{CD}$, $\angle ABC \cong \angle CDA$, and diagonal $\overline{AC}$ is drawn. What is the most direct congruence method that could be used to prove $\triangle ABC$ is congruent to $\triangle CDA$?

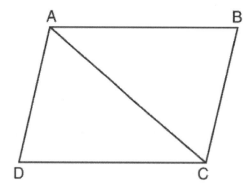

9.6 Prove Triangles Congruent

Model Problem

Given: $\overline{YW}$ is the perpendicular bisector of $\overline{XZ}$.
Prove: $\overline{XY} \cong \overline{ZY}$

Solution:

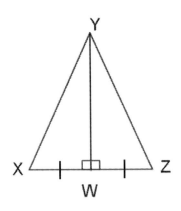

Statements	Reasons
$\overline{YW}$ is the perpendicular bisector of $\overline{XZ}$	Given
$\overline{XW} \cong \overline{ZW}$ (S)	Definition of perpendicular bisector
$\angle YWX$ and $\angle YWZ$ are right angles	Definition of perpendicular bisector
$\angle YWX \cong \angle YWZ$ (A)	Right angles are congruent
$\overline{YW} \cong \overline{YW}$ (S)	Reflexive Property
$\triangle XYW \cong \triangle ZYW$	SAS
$\overline{XY} \cong \overline{ZY}$	CPCTC

Explanation of steps:
(A) Include any given statements that will be used in your proof.
(B) Use the given information to deduce any relevant facts, being sure to mark any congruencies or measures on the diagram.
(C) Continue until you get all three parts of a triangle congruence method.
(D) Conclude that the triangles are congruent by that method.
(E) Use CPCTC to show that any other corresponding parts are congruent, as needed.

This is actually the proof of a theorem called the Perpendicular Bisector Equidistance Theorem, which we will see later in Section 12.2.

Triangles and Congruence — 9.6 Prove Triangles Congruent

Practice Problems

1. Complete the partial proof below for the accompanying diagram by providing reasons for steps 3, 6, 8, and 9.

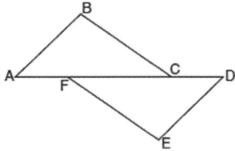

Given: $\overline{AFCD}, \overline{AB} \perp \overline{BC}, \overline{DE} \perp \overline{EF}, \overline{BC} \parallel \overline{FE}, \overline{AB} \cong \overline{DE}$
Prove: $\overline{AC} \cong \overline{FD}$

Statements	Reasons
1. $\overline{AFCD}$	1. Given
2. $\overline{AB} \perp \overline{BC}, \overline{DE} \perp \overline{EF}$	2. Given
3. ∠B and ∠E are right angles	3. _____
4. ∠B ≅ ∠E	4. Right angles are congruent
5. $\overline{BC} \parallel \overline{FE}$	5. Given
6. ∠BCA ≅ ∠EFD	6. _____
7. $\overline{AB} \cong \overline{DE}$	7. Given
8. △ABC ≅ △DEF	8. _____
9. $\overline{AC} \cong \overline{FD}$	9. _____

Triangles and Congruence   9.6 Prove Triangles Congruent

2. Given: $\overline{BE}$ and $\overline{AD}$ intersect at C, $\overline{BC} \cong \overline{EC}$, and $\overline{AC} \cong \overline{DC}$
   Prove: △ABC ≅ △DEC

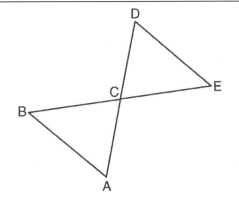

Statements	Reasons

3. Given: $\overline{FH} \cong \overline{FI}$, $\overline{SH} \cong \overline{SI}$
   Prove: ∠H ≅ ∠I

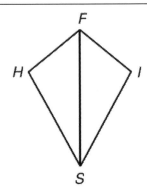

Statements	Reasons

4. Given: $\overline{AD}$ bisects $\overline{BC}$ at E, $\overline{AB} \perp \overline{BC}, \overline{DC} \perp \overline{BC}$
   Prove: $\overline{AB} \cong \overline{DC}$

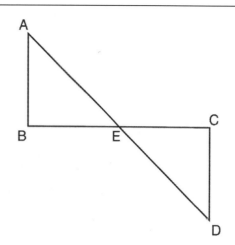

Statements	Reasons

Triangles and Congruence   9.6 Prove Triangles Congruent

5. Given: △ABC, $\overline{BD}$ bisects ∠ABC, $\overline{BD} \perp \overline{AC}$
    Prove: $\overline{AB} \cong \overline{CB}$

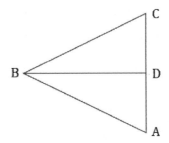

Statements	Reasons

6. Given: $\overline{AFEC}$, $\overline{AF} \cong \overline{EC}$, ∠1 ≅ ∠2, ∠3 ≅ ∠4
    Prove: △ABE ≅ △CDF

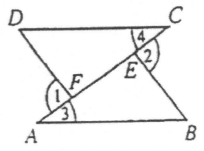

Statements	Reasons

9.7 Overlapping Triangles

MODEL PROBLEM 1: SELECTING A CONGRUENCE METHOD

In the diagram to the right, $\overline{AC} \cong \overline{BD}$ and $\overline{BC} \cong \overline{AD}$.
What is the most direct congruence method that could be used to prove $\triangle ABC \cong \triangle BAD$?

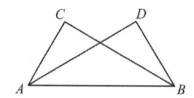

Solution:

(A)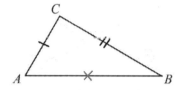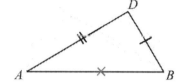

(B)

(C) SSS

Explanation of steps:
(A) Separate out the overlapping triangles and mark the given pairs of congruent parts.
(B) For overlapping triangles, look for any sides or angles that are shared by both triangles. To show that a side or angle is congruent to itself, an *X* is commonly used. [The two triangles share the same side, $\overline{BC}$, which must be congruent to itself by the Reflexive Property.]
(C) Determine which congruence method should be used.

PRACTICE PROBLEMS

1. In the diagram below, if we are given that ∠ACB ≅ ∠DBC and ∠A ≅ ∠D, as shown, what is the most direct congruence method that could be used to prove $\triangle ABC \cong \triangle DCB$?

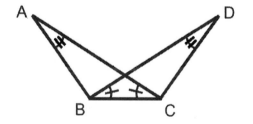

2. In the diagram below, if we are given that $\overline{AC} \cong \overline{AD}$ and $\overline{AE} \cong \overline{AB}$, what is the most direct congruence method that could be used to prove $\triangle ACE \cong \triangle ADB$?

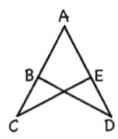

Triangles and Congruence 9.7 Overlapping Triangles

MODEL PROBLEM 2: *WRITING A PROOF*

Given: $\overline{AEFB}, \overline{CE} \cong \overline{DF}, \angle 1 \cong \angle 2, \overline{AE} \cong \overline{BF}$
Prove: $\triangle AFD \cong \triangle BEC$

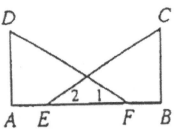

Solution:

Statements		Reasons
$\overline{AEFB}, \overline{CE} \cong \overline{DF}, \angle 1 \cong \angle 2, \overline{AE} \cong \overline{BF}$ (S, A)		Given
$EF = EF$		Reflexive Property
$AE + EF = BF + EF$ $\overline{AF} \cong \overline{BE}$	(S)	Addition Property
$\triangle AFD \cong \triangle BEC$		SAS

Explanation of steps:

(A) Separate the triangles and label the given congruent parts:

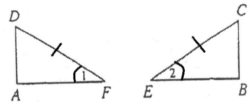

(B) Look for segments or angles that the two overlapping triangles share. *[The corresponding sides $\overline{AF}$ and $\overline{BE}$ share the segment $\overline{EF}$. Since $\overline{AE} \cong \overline{BF}$, adding $\overline{EF}$ to both $\overline{AE}$ and $\overline{BF}$ would give us congruent corresponding sides $\overline{AF}$ and $\overline{BE}$.]*

(C) Prove the congruence of any remaining parts needed to prove the triangles congruent by one of the congruence methods *[$\triangle AFD \cong \triangle BEC$ by SAS]*.

Triangles and Congruence   9.7 Overlapping Triangles

Practice Problems

3. Given: $\overline{DA} \cong \overline{CB}, \overline{DA} \perp \overline{AB}, \overline{CB} \perp \overline{AB}$
   Prove: $\triangle DAB \cong \triangle CBA$

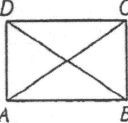

Statements	Reasons

4. Given: $\overline{AB} \cong \overline{CB}, \overline{AD} \cong \overline{CE}$
   Prove: $\angle ADC \cong \angle CEA$

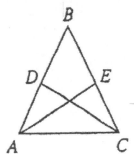

Statements	Reasons

137

Triangles and Congruence 9.7 Overlapping Triangles

5. Given: $\angle C \cong \angle D, \overline{AC} \cong \overline{AD}$
   Prove: $\overline{CE} \cong \overline{DB}$

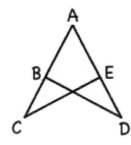

Statements	Reasons

Chapter 10. Triangles and Similarity

10.1 Properties of Similar Triangles

MODEL PROBLEM

In the diagram below, $\triangle ABC \sim \triangle ADE$. If $AD = 8$, $DC = 1$, and $AE = 6$, find EB.

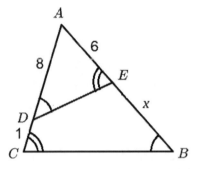

Solution:

(A) $\dfrac{AB}{AD} = \dfrac{AC}{AE}$

(B) $\dfrac{x+6}{8} = \dfrac{9}{6}$

(C) $6(x+6) = 9 \cdot 8$

$6x + 36 = 72$

$6x = 36$

$x = 6$

Explanation of steps:

(A) Write a ratio using corresponding sides in the proper order. [Given $\triangle ABC \sim \triangle ADE$, we know $\dfrac{AB}{AD} = \dfrac{BC}{DE} = \dfrac{AC}{AE}$. However, nothing is given about the lengths of BC and DE, so we can omit the middle ratio.]

(B) Substitute the given lengths. [$AB = EB + AE = x + 6$ and $AC = DC + AD = 1 + 8 = 9$.]

(C) Solve the proportion by cross multiplying.

Practice Problems

1. △ABC ~ △PQR. Find m∠Q.

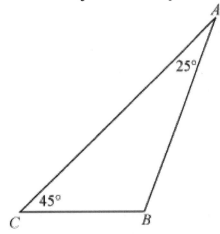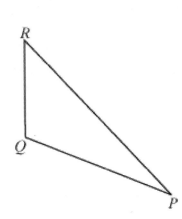

2. If △ABC ~ △ZXY, m∠A = 50, and m∠C = 30, what is m∠X?

3. △ABC ~ △XYZ. Find AC.

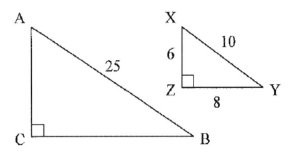

4. △ABC ~ △XYZ. Find XY.

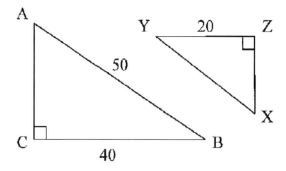

5. In the accompanying diagram, △QRS is similar to △LMN, $RQ = 30$, $QS = 21$, $SR = 27$, and $LN = 7$. What is the length of $\overline{ML}$?

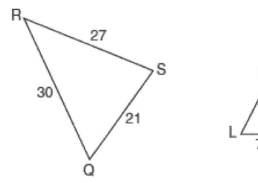

6. The sides of a triangle measure 7, 4, and 9. If the longest side of a similar triangle measures 36, determine and state the length of the shortest side of this triangle.

7. Triangle *ABC* is similar to triangle *DEF*. The lengths of the sides of △ *ABC* are 5, 8, and 11. What is the length of the shortest side of △ *DEF* if its perimeter is 60?

8. The sides of a triangle are 8, 12, and 15. The longest side of a similar triangle is 18. What is the ratio of the perimeter of the smaller triangle to the perimeter of the larger triangle?

10.2 Triangle Similarity Methods

Model Problem

Given: $\overline{AB} \parallel \overline{DE}$, $\overline{AE}$ and $\overline{BD}$ intersect at C, $AB = 20$, $BC = 15$, and $DE = 12$. Find CD.

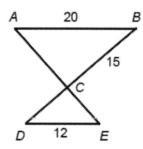

Solution:
(A) $\angle A \cong \angle E$ and $\angle B \cong \angle D$
(B) $\triangle ABC \sim \triangle EDC$
(C) $\dfrac{CD}{15} = \dfrac{12}{20}$
(D) $CD = 9$

Explanation of steps:
(A) When parallel lines are intersected by a transversal, pairs of congruent alternate interior angles are formed. *[It is sufficient to show that two pairs of corresponding angles are congruent, but it is also true that $\angle ACB \cong \angle ECD$ since they are a pair of vertical angles.]*
(B) Triangles are congruent by AA~.
(C) Corresponding sides of similar triangles are in proportion.
(D) Solve for CD.

Practice Problems

1. In the diagram below, $\overline{QR} \parallel \overline{TU}$. Name the pair of similar triangles, using corresponding order. Why are the triangles similar?

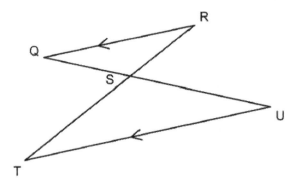

2. In the diagram below, $\overline{AB} \parallel \overline{CD}$. Name the pair of similar triangles, using corresponding order. Why are the triangles similar?

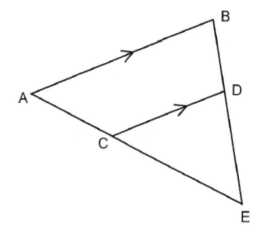

3. In △ ABC and △ DEF, $\frac{AC}{DF} = \frac{CB}{FE}$.
   Which additional information would prove △ ABC ≅ △ DEF?

   (1) AC = DF   (3) ∠ACB ≅ ∠DFE
   (2) CB = FE   (4) ∠BAC ≅ ∠EDF

4. In triangles ABC and DEF, AB = 4, AC = 5, DE = 8, DF = 10, and ∠A ≅ ∠D. What method could be used to prove △ ABC ≅ △ DEF?

5. In the diagram below, $\overline{SQ}$ and $\overline{PR}$ intersect at T, $\overline{PQ}$ is drawn, and $\overline{PS} \parallel \overline{QR}$. What method can be used to prove that △ PST ~ △ RQT?

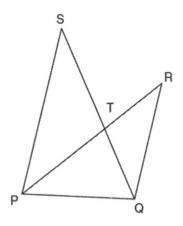

6. For which diagram is the statement △ ABC ≅ △ ADE not always true? (Assume that all lines that appear to be straight are straight lines.)

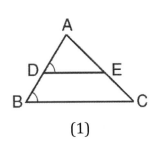

(1)

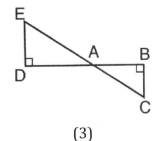

(3)

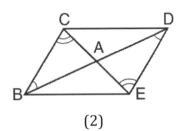

(2)

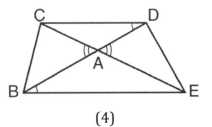

(4)

10.3 Prove Triangles Similar

Model Problem

Given: $\overline{AOD}$ and $\overline{BOC}$, $\dfrac{AO}{OD} = \dfrac{BO}{OC}$

Prove: $\overline{AB} \parallel \overline{CD}$

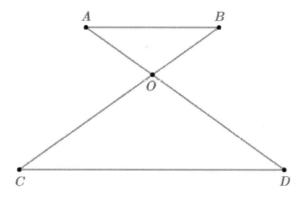

Solution:

Statements		Reasons
$\dfrac{AO}{DO} = \dfrac{OB}{OC}$	(S, S)	Given
$\angle AOB \cong \angle DOC$	(A)	Vertical $\angle$'s are $\cong$
$\triangle AOB \sim \triangle DOC$		SAS~
$\angle A \cong \angle D$ [or $\angle B \cong \angle C$]		CASTC
$\overline{AB} \parallel \overline{CD}$		Alternate Interior Angles Converse

Explanation of steps:
(A) Use the given information, and any other facts derived from the diagram, to prove two triangles are similar. *[We are given that two pairs of corresponding sides are proportional. Without any additional information about the third pair of sides, we cannot use the SSS~ method. However, the included angles between these two pairs of sides are vertical angles, which are always congruent. Therefore, the triangles are similar by SAS~.]*
(B) When we know the triangles are similar, any other pairs of corresponding angles are congruent by CASTC (or any other pairs of corresponding sides are also proportional by CSSTP). *[$\angle A \cong \angle D$ by CASTC. Or, we could have used $\angle B \cong \angle C$ by CASTC instead.]*
(C) Complete the proof. *[Since we have alternate interior angles that are congruent, it must be true that $\overline{AB} \parallel \overline{CD}$.]*

Practice Problems

1. Given: △ABC and △ADE with ∠ACB ≅ ∠AED
   Prove: △ABC ~ △ADE.

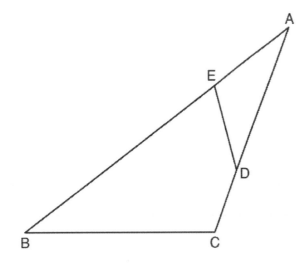

2. Given: $\overline{BFCE}$, $\overline{AB} \perp \overline{BE}$, $\overline{DE} \perp \overline{BE}$, and ∠BFD ≅ ∠ECA
   Prove: △ABC ~ △DEF.

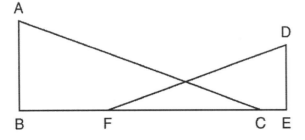

3. Given: △PQR and △STU with right ∠Q and right ∠T,
PQ = 6, QR = 8, ST = 3, and TU = 4.
Prove: ∠R ≅ ∠U

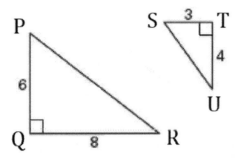

10.4 Triangle Angle Bisector Theorem

Model Problem

In the diagram to the right, $\overline{LK}$ bisects ∠L. Find x.

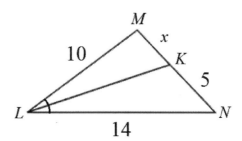

Solution:
$$\frac{x}{5} = \frac{10}{14}$$
$$14x = 50$$
$$x \approx 3.57$$

Explanation of steps:
The bisector of an angle of a triangle splits the opposite side into segments that are proportional to the adjacent sides. $\left[\frac{MK}{NK} = \frac{ML}{NL}\right]$

Practice Problems

1. In the diagram below, an angle bisector of the triangle is shown. Find x.

   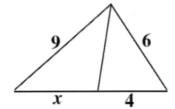

2. In the diagram below, an angle bisector of the triangle is shown. Find x.

   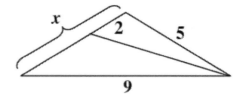

3. In the diagram below, $\overline{TB}$ bisects $\angle ATC$, $TA = 21$, $TC = 24$, and $AC = 30$. Find AB.

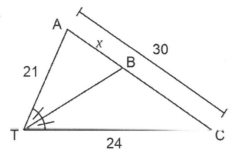

4. In $\triangle ABC$ below, $\overline{AD}$ bisects $\angle BAC$. $AB = x$, $BD = x - 2$, $DC = x + 1$, and $AC = x + 4$.
Find the perimeter of $\triangle ABC$.

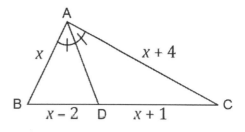

5. In $\triangle DEF$, $\overline{DG}$ bisects $\angle D$, $m\angle E = 90°$, $DE = 6$ and $EF = 8$. Find DG.

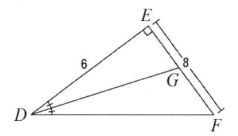

10.5 Side Splitter Theorem

Model Problem

In the diagram below, $\overline{PQ} \parallel \overline{TR}$. Find x.

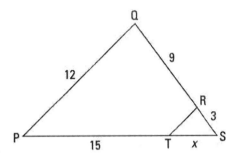

Solution:
$$\frac{x}{15} = \frac{3}{9}$$
$$9x = 45$$
$$x = 5$$

Explanation of steps:
A line parallel to one side of a triangle divides the other two sides proportionally. $\left[\frac{ST}{TP} = \frac{SR}{RQ}\right]$

Practice Problems

1. In the diagram below of $\triangle ACT$, $\overleftrightarrow{BE} \parallel \overline{AT}$.

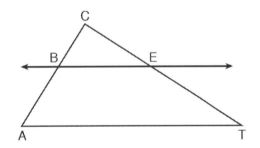

If $CB = 3$, $CA = 10$, and $CE = 6$, what is the length of $\overline{ET}$?

2. In the diagram below, $\overline{AE} \parallel \overline{BD}$. Find x.

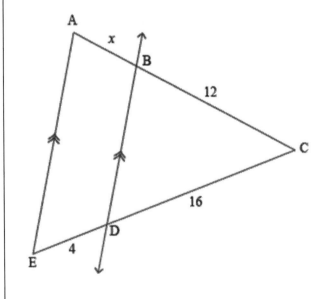

3. In the diagram of △ ABC below, $\overline{DE} \parallel \overline{BC}$, $AD = 3$, $DB = 2$, and $DE = 6$.

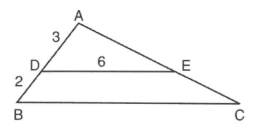

What is the length of $\overline{BC}$?

4. In the diagram of △ ABC shown below, $\overline{DE} \parallel \overline{BC}$.

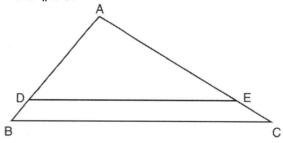

If $AB = 10$, $AD = 8$, and $AE = 12$, what is the length of $\overline{EC}$?

5. In the diagram below of △ ACD, E is a point on $\overline{AD}$ and B is a point on $\overline{AC}$, such that $\overline{EB} \parallel \overline{DC}$. If $AE = 3$, $ED = 6$, and $DC = 15$, find the length of $\overline{EB}$.

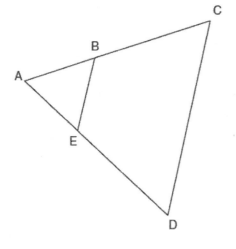

6. In the diagram below of △ ABC, D is a point on $\overline{AB}$, E is a point on $\overline{BC}$, $\overline{AC} \parallel \overline{DE}$, $CE = 25$ inches, $AD = 18$ inches, and $DB = 12$ inches. Find EB, to the *nearest tenth of an inch*.

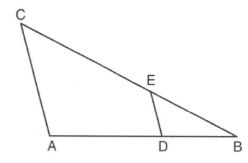

7. In the diagram, $\overline{NR} \parallel \overline{PQ}$. $MQ = 42$, $MN = 13$, and $NP = 8$. Find RQ and MR.

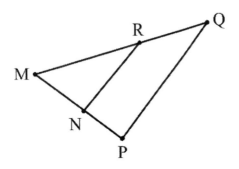

8. In the diagram of $\triangle ABC$ below, $\overline{DE} \parallel \overline{AB}$.

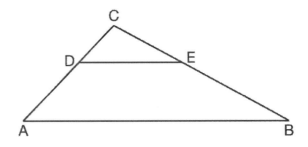

If $CD = 4$, $CA = 10$, $CE = x + 2$, and $EB = 4x - 7$, what is the length of $\overline{CE}$?

9. In the diagram below of $\triangle ADE$, B is a point on $\overline{AE}$ and C is a point on $\overline{AD}$ such that $\overline{BC} \parallel \overline{ED}$, $AC = x - 3$, $BE = 20$, $AB = 16$, and $AD = 2x + 2$. Find the length of $\overline{AC}$.

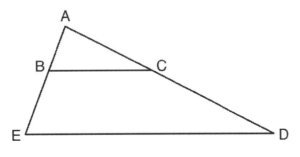

10. In triangle BWL: $\overline{WE}$ is the bisector of $\angle W$, with E on $\overline{BL}$. $\overline{EO}$ is parallel to $\overline{BW}$, with O on $\overline{WL}$. $WO = 6$ and $OL = 9$. Find BW.

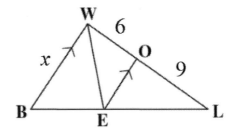

11. The diagram below shows a section of a city map. Willow Road, Pine Street, and Oak Drive are parallel and 5 miles apart. Ridge Road is perpendicular to the three parallel streets. The distance between the intersection of Ridge Road and Pine Street and where the railroad tracks cross Pine Street is 12 miles. What is the distance between the intersection of Ridge Road and Oak Drive and where the railroad tracks cross Oak Drive?

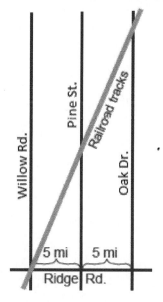

10.6 Triangle Midsegment Theorem

Model Problem

In $\triangle ABC$, $\overline{DE}$ and $\overline{EF}$ are midsegments, $AB = 53$, $DE = 25$, and m$\angle B = 92°$.
Find EF, FC, and m$\angle CFE$.

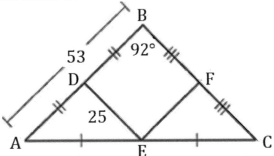

Solution:
(A) $EF = \frac{1}{2} AB = 26.5$
(B) $BC = 2(25) = 50$, so $FC = \frac{1}{2}(50) = 25$
(C) $\overline{EF} \parallel \overline{AB}$, so m$\angle CFE = 92°$

Explanation of steps:
(A) A midsegment is half the length of the parallel side.
(B) The side parallel to a midsegment is twice the length of the midsegment. *[Since F is the midpoint of $\overline{BC}$, $FC = \frac{1}{2} BC$.]*
(C) A midsegment is parallel to the third side. *[$\angle B$ and $\angle CFE$ are corresponding angles formed by parallel lines intersected by transversal $\overline{BC}$, so they are congruent.]*

Triangles and Similarity — 10.6 Triangle Midsegment Theorem

Practice Problems

1. Find x.

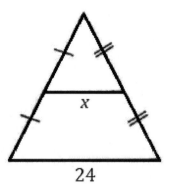

2. In the diagram of △ ABC below, D is the midpoint of $\overline{AB}$ and E is the midpoint of $\overline{BC}$. $DE = 2x + 2$ and $AC = 5x$. Find DE and AC.

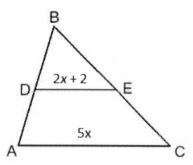

3. In the diagram below, the vertices of △ DEF are the midpoints of the sides of equilateral triangle ABC, and the perimeter of △ ABC is 36 cm.

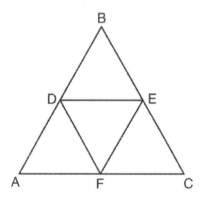

What is the length, in cm, of $\overline{EF}$?

4. In the diagram of △ ABC shown below, D is the midpoint of $\overline{AB}$, E is the midpoint of $\overline{BC}$, and F is the midpoint of $\overline{AC}$.

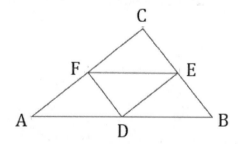

If $AB = 20$, $BC = 12$, and $AC = 16$, what is the perimeter of trapezoid ABEF?

5. In △ABC shown below, L is the midpoint of $\overline{BC}$, M is the midpoint of $\overline{AB}$, and N is the midpoint of $\overline{AC}$.

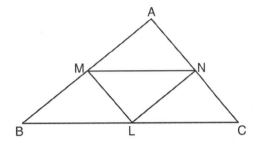

If $MN = 8$, $ML = 5$, and $NL = 6$, what is the perimeter of trapezoid BMNC?

6. In isosceles triangle RST shown below, $\overline{RS} \cong \overline{RT}$, M and N are midpoints of $\overline{RS}$ and $\overline{RT}$, respectively, and $\overline{MN}$ is drawn. If $MN = 3.5$ and the perimeter of △RST is 25, find the length of $\overline{NT}$.

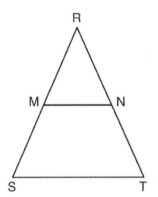

7. In the diagram below, find PR, ST, m∠STP, and m∠SUR.

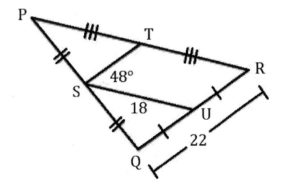

Chapter 11. Points of Concurrency

11.1 Incenter and Circumcenter

Model Problem

$\overline{BP}$ and $\overline{CP}$ are angle bisectors of $\triangle ABC$, $\overline{PT} \perp \overline{BC}$, m∠A = 40°, m∠ABP = 36°, and $PT = 9$.

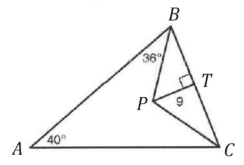

a) Find the distance from P to $\overline{AC}$.
b) Find the area of the circle that can be inscribed in this triangle.
c) Find m∠PCA.

Solutions:
(A) 9
(B) $A = \pi r^2 = \pi \cdot 9^2 = 81\pi$
(C) m∠ABC = 2 · 36° = 72°
    m∠ACB = 180° − (40° + 72°) = 68°
    m∠PCA = $\frac{1}{2}$ m∠ACB = $\frac{1}{2}$ · 68° = 34°

Explanation of steps:
(A) The incenter of a triangle is the point of intersection of its angle bisectors *[P is the incenter of △ ABC]*. The incenter is equidistant from the three sides of the triangle *[The distance from P to $\overline{AC}$ is equal to PT = 9]*.
(B) The distance from the incenter to any side of the triangle is the inradius, which is the radius of the inscribed circle. *[r = PT = 9]*.
(C) An angle bisector divides an angle in half *[m∠ABC = 2 · m∠ABP = 72°]*. The angles of a triangle add to 180° *[m∠ACB = 68°, so m∠PCA = 34°]*.

Practice Problems

1. In the diagram below of △ABC, $\overline{CD}$ is the bisector of ∠BCA, $\overline{AE}$ is the bisector of ∠CAB, and $\overline{BG}$ is drawn.

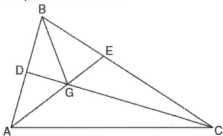

Which statement must be true?

(1) DG = EG   (3) ∠AEB ≅ ∠AEC
(2) AG = BG   (4) ∠DBG ≅ ∠EBG

2. In the diagram below, point B is the incenter of △FEC, and $\overline{EBR}$, $\overline{CBD}$, and $\overline{FB}$ are drawn.

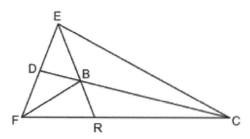

If m∠FEC = 84 and m∠ECF = 28, find m∠BRC.

3. $\overline{PT}$ and $\overline{RT}$ are angle bisectors of △PQR and $\overline{TS} \perp \overline{PR}$. m∠Q = 39°, m∠TPR = 18°, and TS = 3.

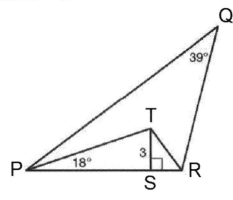

a) Find the circumference of the circle that can be inscribed in △PQR.

b) Find m∠TRP.

4. $\overline{DG}, \overline{EG}$, and $\overline{FG}$ are the perpendicular bisectors of △ABC. DC = 15, BF = 16, EC = 17, and BG = 19.

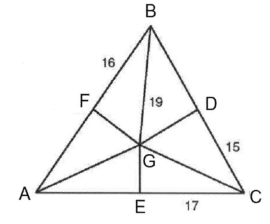

a) Find BD, AF, and AE.

b) Find AG and CG.

c) Find the area of the circle that can be circumscribed about △ABC.

5. P is the incenter of △LMN. MP = 13.4, MQ = 9.4, and $\overline{PQ} \perp \overline{MN}$. Find the area of the circle that can be inscribed in △LMN, to the *nearest tenth of a square unit*.

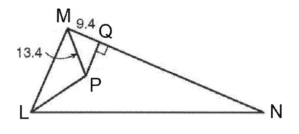

6. C is the circumcenter of △PQR. $CZ = \sqrt{53}$ and RZ = 26. Find the circumference of the circle that can be circumscribed about △PQR.

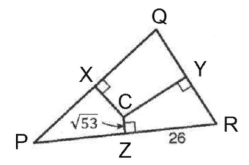

11.2 Orthocenter and Centroid

MODEL PROBLEM

In the diagram of △ ABC below, Jose found centroid P by constructing the three medians. He measured $\overline{CF}$ and found it to be 6 inches.

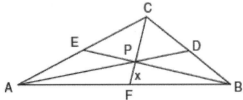

If $PF = x$, write an equation that could be used to find x. Solve for x.

Solution:
(A) $x + 2x = 6$
(B) $3x = 6$
$x = 2$

Explanation of steps:
(A) The centroid of a triangle *[P]* divides the median $\overline{[CF]}$ in a ratio of 2 : 1, with the larger segment $\overline{[CP]}$ at the vertex.
[Since PF + CP = CF = 6, and CP = 2 · PF, we can write the equation, x + 2x = 6.]
(B) Solve for x. *[PF = 2 and CP = 4.]*

PRACTICE PROBLEMS

1. Name the point of concurrency for each of the following in a triangle: a) perpendicular bisectors b) medians c) altitudes d) angle bisectors	2. In which triangle do the three altitudes intersect outside the triangle? (1) a right triangle (2) an acute triangle (3) an obtuse triangle (4) an equilateral triangle

3. For a triangle, which two points of concurrence could be located outside the triangle?

   (1) incenter and centroid
   (2) centroid and orthocenter
   (3) incenter and circumcenter
   (4) circumcenter and orthocenter

4. In the diagram below of $\triangle ABC$, $\overline{AE} \cong \overline{BE}$, $\overline{AF} \cong \overline{CF}$, and $\overline{CD} \cong \overline{BD}$.

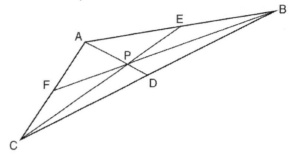

   Point P must be the

   (1) centroid   (3) incenter
   (2) circumcenter (4) orthocenter

5. P is the centroid of $\triangle QRS$, $PT = 8$, and $ST = 24$. Find:

   a) QS

   b) PR

   c) RT

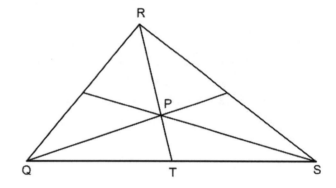

6. In triangle SAT below, P is a centroid, $PR = 12$, $PT = 28$, $AR = 20$, and $AY = 27$. Find SP, TM, AT, and PY.

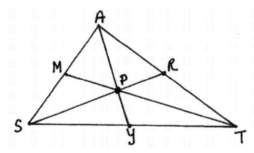

SP = _____

TM = _____

AT = _____

PY = _____

7. In the diagram below of △ACE, medians $\overline{AD}, \overline{EB}$, and $\overline{CF}$ intersect at G. The length of $\overline{FG}$ is 12 cm.

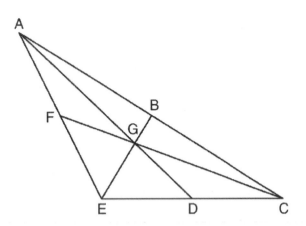

What is the length, in cm, of $\overline{GC}$?

8. In the diagram of △ABC below, medians $\overline{AD}$ and $\overline{BE}$ intersect at point F.

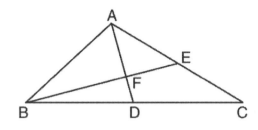

If $AF = 6$, what is the length of $\overline{FD}$?

9. In the diagram below of △MAR, medians $\overline{MN}, \overline{AT}$, and $\overline{RH}$ intersect at O.

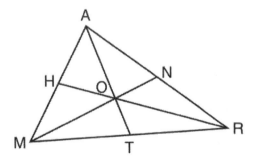

If $TO = 10$, what is the length of $\overline{TA}$?

10. In △ABC shown below, P is the centroid and $BF = 18$.

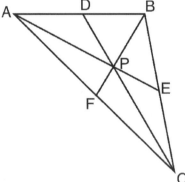

What is the length of $\overline{BP}$?

11. In the diagram below of △ABC, medians $\overline{AD}, \overline{BE}$, and $\overline{CF}$ intersect at G.

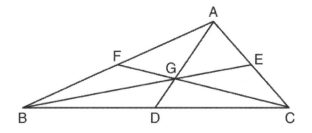

If $CF = 24$, what is the length of $\overline{FG}$?

12. In △ABC shown below, medians $\overline{AD}, \overline{BE}$, and $\overline{CF}$ intersect at point R.

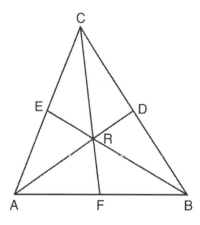

If $CR = 24$ and $RF = 2x - 6$, what is the value of x?

13. In the diagram below, point P is the centroid of △ABC.

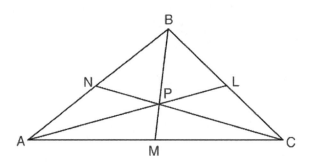

If $PM = 2x + 5$ and $BP = 7x + 4$, what is the length of $\overline{PM}$?

14. In the diagram below, $\overline{QM}$ is a median of triangle PQR and point C is the centroid of triangle PQR.

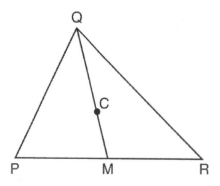

If $QC = 5x$ and $CM = x + 12$, determine and state the length of $\overline{QM}$.

Chapter 12. Right Triangles

12.1 Congruent Right Triangles

Model Problem

Given: Triangles ABD and ACD with right angles at B and C, $\overline{AB} \cong \overline{CD}$.
Prove: $\overline{AC} \cong \overline{BD}$.

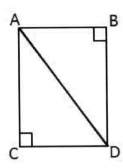

Solution:
1. $\angle B$ and $\angle C$ are right angles (Given)
2. $\overline{AD} \cong \overline{AD}$ (Reflexive Property)
3. $\overline{AB} \cong \overline{CD}$ (Given)
4. $\triangle ABD \cong \triangle DCA$ (HL)
5. $\overline{AC} \cong \overline{BD}$ (CPCTC)

Practice Problems

1. Four pairs of triangles are shown below, with congruent corresponding parts labeled in each pair. Using only the information given in the diagrams, which pair of triangles can *not* be proven congruent?

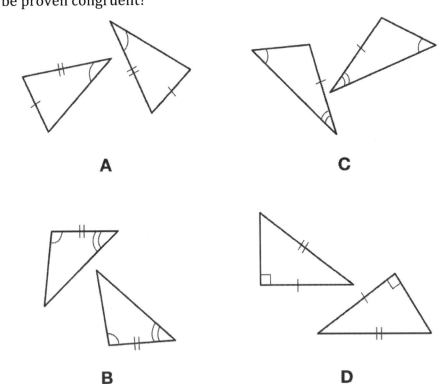

2. For these triangles, select the triangle congruence statement and the congruence method that supports it.

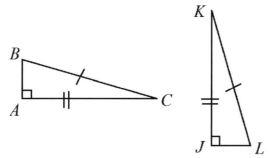

(1) △ ABC ≅ △ JKL, HL (3) △ ABC ≅ △ JLK, HL
(2) △ ABC ≅ △ JKL, SAS (4) △ ABC ≅ △ JLK, SAS

3. Given: Right triangles MAT and HTA, $\overline{MT} \cong \overline{AH}$
   Prove: ∠M ≅ ∠H

4. Given: $\overline{CA} \perp \overline{AB}, \overline{ED} \perp \overline{DF}, \overline{CE} \cong \overline{BF}$, and $\overline{AB} \cong \overline{ED}$.
   Prove: $\overline{AC} \cong \overline{DF}$

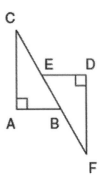

12.2 Equidistance Theorems

Model Problem

In the diagram of △ABC below, $AN = 4$ and $AE = 2\sqrt{3}$. Find NF.

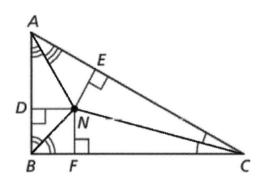

Solution:

(A) $(NE)^2 + (2\sqrt{3})^2 = 4^2$
$(NE)^2 + 12 = 16$
$(NE)^2 = 4$
$NE = \sqrt{4} = 2$

(B) $NF = NE = 2$

Explanation of steps:

(A) Recognizing that we are given the lengths of two sides of a right triangle [△ AEN], we can find the length of the third side [NE] by the Pythagorean Theorem.

(B) The incenter of a triangle [N is the point of concurrency of the three angle bisectors of △ ABC] is equidistant to the three sides of the triangle [so, $NE = NF = ND$].

Practice Problems

1. $\overrightarrow{FH}$ is the bisector of ∠EFG. $EF = 9$ and $EH = 12$. Find HG.

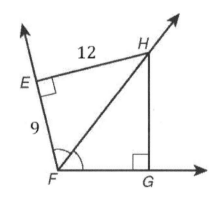

2. $\overline{TQ} \perp \overline{RQ}, \overline{TS} \perp \overline{RS}, TQ = 7, TS = 7$, and m∠QRT = 26°. Find m∠SRT.

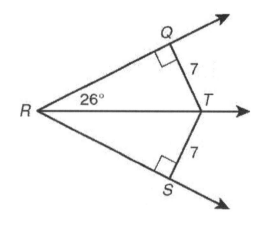

3. $\overleftrightarrow{SQ}$ is the perpendicular bisector of $\overline{PR}$ and $PS = 15$. Find RS.

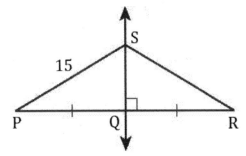

If $m\angle P = 30°$, find $m\angle RSQ$.

4. Find x.

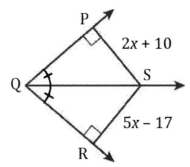

5. Find x.

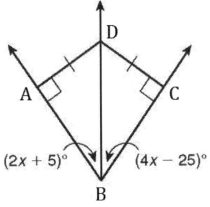

6. Find x.

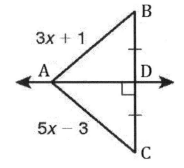

If $m\angle BAC = (41x)°$, find $m\angle B$.

7. Name two line segments that are congruent to $\overline{PQ}$.

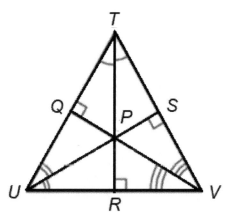

8. Name two line segments that are congruent to $\overline{PA}$.

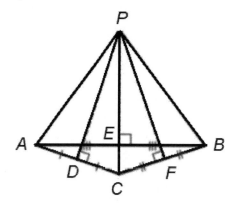

12.3 Geometric Mean Theorems

MODEL PROBLEM

In the diagram below, $\overline{BE}$ bisects $\angle ABC$.
a) Find DC.   b) Find BC.   c) Find DE to the *nearest tenth*.

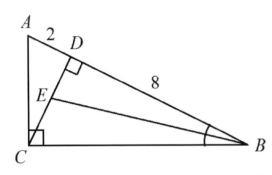

Solution:
(A) By the Altitude Rule,
$(DC)^2 = 2 \cdot 8$
$(DC)^2 = 16$
$DC = 4$

(B) By the Legs Rule,
$(BC)^2 = 8(8 + 2)$
$(BC)^2 = 80$
$BC = \sqrt{80} = 4\sqrt{5}$

(C) By the Triangle Angle Bisector Theorem,

$\dfrac{DE}{EC} = \dfrac{BD}{BC} \qquad \dfrac{x}{4-x} = \dfrac{8}{4\sqrt{5}}$

$4\sqrt{5} \cdot x = 8(4 - x)$
$4\sqrt{5} \cdot x = 32 - 8x$
$4\sqrt{5} \cdot x + 8x = 32$

$(4\sqrt{5} + 8)x = 32$

$x = \dfrac{32}{4\sqrt{5} + 8} \approx 1.9$

Explanation of steps:
(A) $\overline{DC}$ is the altitude of $\triangle ACB$, so it is the geometric mean of AD and BD. $[(DC)^2 = AD \cdot BD]$
(B) $\overline{BC}$ is a leg of $\triangle ACB$, so it is the geometric mean of its projection onto the hypotenuse, BD, and the hypotenuse, AB. $[(BC)^2 = BD \cdot AB]$
  (Note: alternatively, the Pythagorean Theorem could have been used on $\triangle CDB$.)
(C) The bisector of an angle of a triangle splits the opposite side into segments that are proportional to the adjacent sides.

Practice Problems

1. In △ABC, m∠ACB = 90° and altitude $\overline{CD}$ is drawn. If $AD = 3$ and $DB = 9$, find CD in simplest radical form.

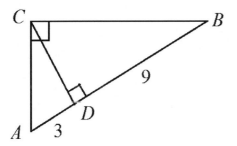

2. In △ABC, m∠ACB = 90° and altitude $\overline{CD}$ is drawn. If $AD = 3$ and $DB = 9$, find AC.

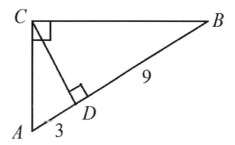

3. In △ABC, m∠ABC = 90° and altitude $\overline{BD}$ is drawn. If $CD = 7$ and $CA = 16$, find BD in simplest radical form.

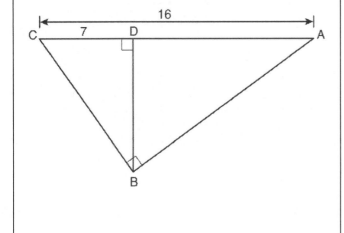

4. In △ABC, m∠BAC = 90° and altitude $\overline{AD}$ is drawn. If $BD = 2$ and $DC = 10$, find AB in simplest radical form.

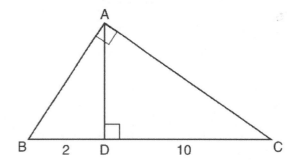

5. In △ABC, m∠ACB = 90° and altitude $\overline{CD}$ is drawn. If AD = 3 and DB = 4, find CB in simplest radical form.

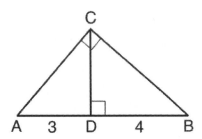

6. In △PQR, m∠PQR = 90° and altitude $\overline{QM}$ is drawn. If PM = 8 and MR = 18, find QM.

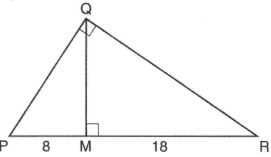

7. In △ABC, m∠ACB = 90° and altitude $\overline{CD}$ is drawn. If AD = 5 and BC = 6, find BD.

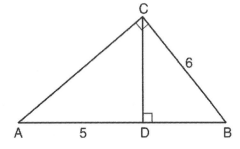

8. In △ABC, m∠ACB = 90° and altitude $\overline{CD}$ is drawn. If AD = 21 and BC = 10, find BD.

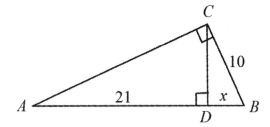

9. In △ABC, ∠ABC is a right angle, altitude BD = 4 meters, and $\overline{DC}$ is 6 meters longer than $\overline{AD}$. Find the length of base $\overline{AC}$ in meters.

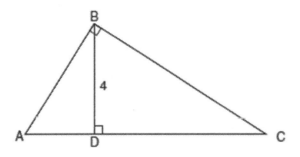

10. Find x, y, and z.

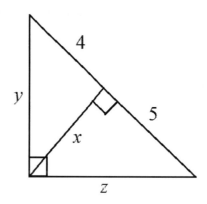

11. Find x, y, and z.

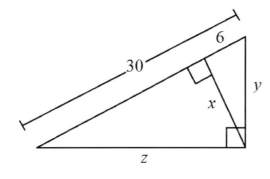

12. In the 30-60-90 triangle below, an altitude is drawn to the hypotenuse. What is the length of this altitude?

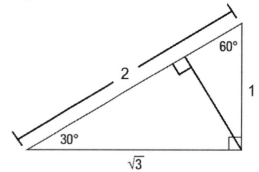

Chapter 13. Trigonometry

13.1 Trigonometric Ratios

Model Problem

For the right triangle in the diagram, find, to the *nearest thousandth*, sin A and tan B.

Solution:

    (A) (B) (C)

$$\sin A = \frac{opp}{hyp} = \frac{5}{13} \approx 0.385$$

$$\tan B = \frac{opp}{adj} = \frac{12}{5} = 2.4$$

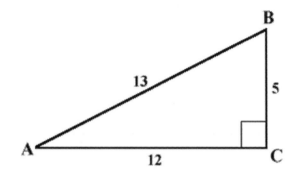

Explanation of steps:
(A) Write the correct trigonometric ratio.
(B) Substitute the lengths of the appropriate sides.
(C) Divide, and round if necessary.

Practice Problems

1. For the right triangle below, find sin A, cos A, and tan A. Leave in fraction form.

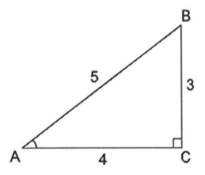

2. For the right triangle below, find sin B, cos B, and tan B. Leave in fraction form.

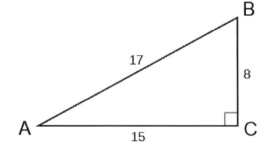

3. For the right triangle below, state two trigonometric functions that are equal to the ratio, $\frac{40}{41}$.

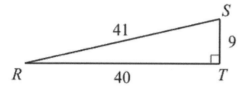

4. For right triangle ABC below, find the tangent of ∠B, to the *nearest thousandth*.

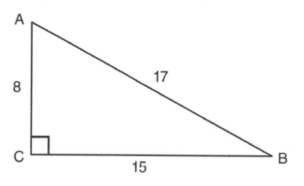

5. For the right triangle shown below, find sin x, to the *nearest thousandth*.

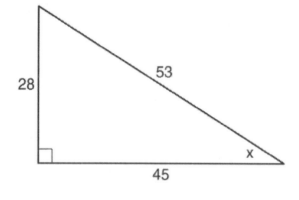

6. In right triangle ABC shown below, what is the value of cos A?

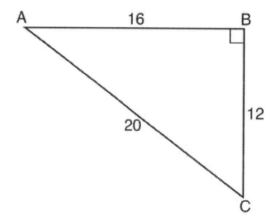

7. In △LMN, m∠M = 90°, LM = 7, MN = 24, and LN = 25. What is the sine of ∠N written as a ratio?

8. In △PQR, m∠PQR = 90°, PQ = 20, QR = 21, and PR = 29. What is the tangent of ∠QPR written as a ratio?

9. In △ABC, shown below, m∠C = 90°.
   What is the sine of ∠B written as a ratio in lowest terms?

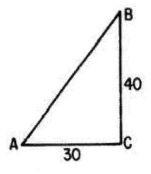

13.2 Use Trigonometry to Find a Side

Model Problem

Given the measures shown in the diagram below, find a to the *nearest tenth of a foot*.

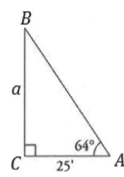

Solution:

(A) $\tan A = \dfrac{opposite}{adjacent}$

(B) $\tan 64° = \dfrac{a}{25}$

(C) $25 \tan 64° = a$

(D) $51.3 \approx a$

Explanation of steps:
(A) Select the appropriate trig ratio for the given and needed measures *[since we are given the adjacent leg and need to find the opposite leg, use the tangent ratio]*.
(B) Substitute the given measures and the variable for the unknown.
(C) Solve for (isolate) the variable *[multiply both sides by 25]*.
(D) Use a calculator to find the answer *[enter 25 tan(64) using the [TAN] key]*.

Practice Problems

1. The hypotenuse of a right triangle measures 10 inches and one angle measures 31°. Which equation could be used to find the length of the side opposite the 31° angle?

   (1) $\sin 31° = \dfrac{10}{x}$

   (2) $\cos 31° = \dfrac{x}{10}$

   (3) $\sin 31° = \dfrac{x}{10}$

   (4) $\tan 31° = \dfrac{x}{10}$

2. As shown in the accompanying diagram, the diagonal of the rectangle is 10 inches long. The diagonal makes a 15° angle with the longer side of the rectangle. What is the width, w, of the rectangle to the *nearest tenth of an inch*?

   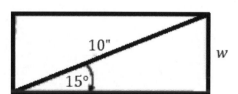

3. The accompanying diagram shows a ramp 30 feet long leaning against a wall at a construction site. If the ramp forms an angle of 32° with the ground, how high above the ground, to the *nearest tenth of a foot*, is the top of the ramp?

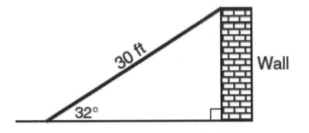

4. Find, to the *nearest tenth of a foot*, the height of the tree represented below.

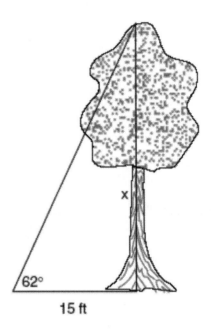

5. An 8-foot rope is tied from the top of a pole to a stake in the ground, as shown in the diagram. If the rope forms a 57° angle with the ground, find the height of the pole, to the *nearest tenth of a foot*.

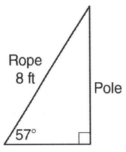

6. A tree casts a 25-foot shadow, as shown in the diagram. If the angle of elevation from the tip of the shadow to the top of the tree is 32°, find the height of the tree, to the *nearest tenth of a foot*.

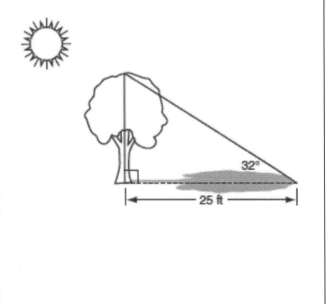

7. A 5-foot ladder leans against a wall and makes an angle of 65° with the ground, as shown below. Find, to the *nearest tenth of a foot*, the distance from the wall to the base of the ladder.

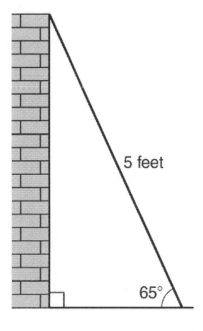

8. A metal pipe holds up a 9-foot fence, as shown below. The pipe makes an angle of 48° with the ground. Find, to the *nearest foot*, the length of the pipe.

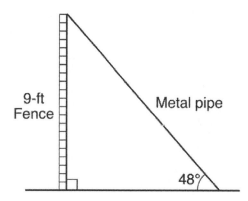

9. An airplane is climbing at an angle of 11° with the ground. Find, to the *nearest foot*, the ground distance the airplane has traveled when it has attained an altitude of 400 feet.

10. Find the area of the triangle below, to the *nearest tenth of a square foot*.

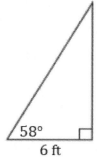

11. A stake is driven into the ground away from the base of a 50-foot pole, as shown below. A wire connects the stake on the ground to the top of the pole at an angle of elevation of 52°.

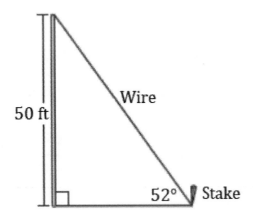

What is the distance from the base of the pole to the stake, to the *nearest foot*?

What is the length of the wire, to the *nearest foot*?

12. A hot-air balloon is tied to the ground with two ropes, as shown below. One rope makes a right angle with the ground. The other rope forms an angle of 50° with the ground.

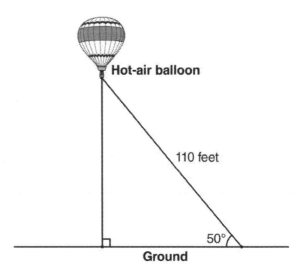

Find the distance, to the *nearest foot*, from the balloon to the ground.

Find the distance, to the *nearest foot*, on the ground between the two ropes.

13. A 10-foot ladder is to be placed against the side of a building. The base of the ladder must be placed at an angle of 72° with the level ground for a secure footing.

    Find, to the *nearest inch*, how far the base of the ladder should be from the side of the building.

    Find how far up the side of the building the ladder will reach.

14. A lighthouse is built on the edge of a cliff near the ocean, as shown in the diagram. From a boat located 200 feet from the base of the cliff, the angle of elevation to the top of the cliff is 18° and the angle of elevation to the top of the lighthouse is 28°.

    What is the height of the lighthouse, x, to the *nearest tenth of a foot*?

    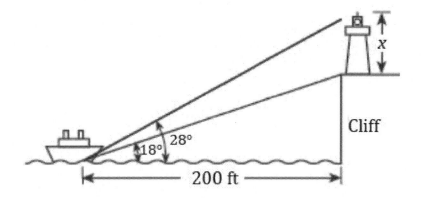

13.3 Use Trigonometry to Find an Angle

Model Problem

In the right triangle below, the lengths of the legs are 9 and 10 as shown. Find the measure of angle *x* to the *nearest degree*.

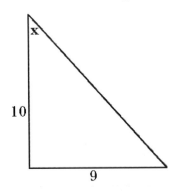

Solution:
(A) $\tan x = \frac{opposite}{adjacent}$

(B) $\tan x = \frac{9}{10}$

(C) $x = \tan^{-1}\left(\frac{9}{10}\right)$

(D) $x \approx 42°$

Explanation of steps:
(A) Select the appropriate trig ratio *[the opposite and adjacent legs are given, so use tan]*.
(B) Substitute the given measures.
(C) Since the angle is the unknown, isolate the variable by taking the inverse trig function of both sides. Think of a trig function and its inverse as "canceling out."
(D) Use the inverse trig function on the calculator. Round your answer as specified.
  [Press [2nd][TAN⁻¹][9][÷][1][0][)][ENTER].]

Practice Problems

1. What is the measure of the angle, *x*, to the *nearest tenth of a degree*?

   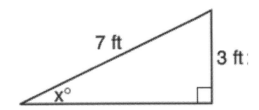

2. What is the measure of ∠A, to the *nearest tenth of a degree*?

   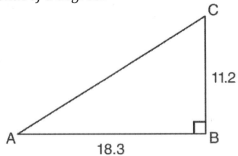

3. As shown below, a 50-foot wire is stretched from the top of a 30-foot antenna to the ground. Find, to the *nearest degree,* the measure of the angle that the wire makes with the ground.

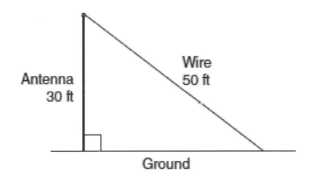

4. An 8-foot pole, standing perpendicular to the ground, holds up a tent. A side of the tent is 12 feet long, as shown below. Find the measure of angle *A*, to the *nearest degree.*

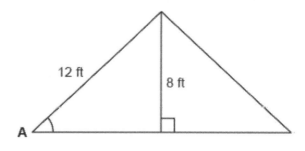

5. A 28-foot ladder leans against a wall. The bottom of the ladder is 6 feet from the base of the wall. Find the measure of the angle formed by the ladder and the ground, to the *nearest degree.*

6. In right $\triangle ABC$, m$\angle C = 90°$, $BC = 10$ and $AB = 16$. Find, to the *nearest tenth of a degree*, the measures of the two acute angles, $\angle A$ and $\angle B$.

183

7. A person standing on level ground is 2,000 feet away from the base of a 420-foot-tall building. Find, to the *nearest degree*, the measure of the angle of elevation to the top of the building from the point on the ground where the person is standing.

8. A person standing on level ground is 1,000 feet away from the base of a 350-foot-tall building. Find, to the *nearest degree*, the measure of the angle of elevation to the top of the building from the point on the ground where the person is standing.

9. The diagram shows a flagpole that stands on level ground. Two cables, r and s, are attached to the pole at a point 16 feet above the ground. The combined length of the two cables is 50 feet. If cable r is attached to the ground 12 feet from the base of the pole, what is the measure of the angle, x, to the *nearest degree*, that cable s makes with the ground?

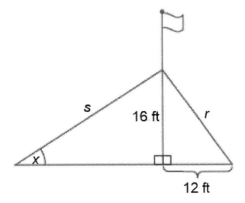

10. A trapezoid is shown below. Find m∠x, to the *nearest tenth of a degree*.

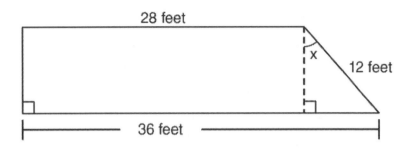

11. The base of a 15-foot ladder rests on the ground 4 feet from a 6-foot fence. The ladder leans so that it touches the top of the fence and the side of a brick wall.

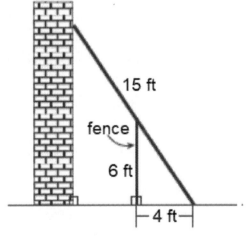

What angle, to the *nearest degree*, does the ladder make with the ground?

How far up the wall does the top of the ladder reach, to the *nearest foot*?

13.4 Special Triangles

MODEL PROBLEM

A straight post leaning against a wall makes a 45° angle with the ground. The base of the post is 3.25 meters from the building. What is the length of the post, to the *nearest tenth of a meter*?

Solution:
$$3.25 \times \sqrt{2} \approx 4.6 \text{ meters}$$

Explanation:
The post is the hypotenuse of a 45-45-90 triangle as it leans against the building. Therefore, if one leg, x, is 3.25 meters, then the hypotenuse can be calculated as $x\sqrt{2} = 3.25 \times \sqrt{2}$.

PRACTICE PROBLEMS

1. In the right triangle shown below, what is the value of x to the *nearest whole number*?

   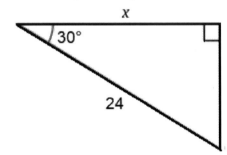

2. The perimeter of a square is 20 inches. Find the length of a diagonal.

3. The altitude of an equilateral triangle is $2\sqrt{3}$ units. Find the length of a side.

13.5 Cofunctions

MODEL PROBLEM

If $\sin 2x = \cos 3x$, find the value of x.

Solution:
$2x + 3x = 90$
$5x = 90$
$x = 18$

Explanation:
Cofunctions of complementary angles are equal. If two angles are complementary, their measures add to 90°.

PRACTICE PROBLEMS

1. If $\sin 25° = \cos B$, find m∠B.

2. If $\cos 72° = \sin x$, find the number of degrees in the measure of acute angle x.

3. If $\sin(x + 15)° = \cos(x - 5)°$, find x.

4. If $\cos(2x - 1)° = \sin(3x + 6)°$, find x.

5. If $\sin(x+20)° = \cos x$, find x.

6. If $\sin(x-3)° = \cos(2x+6)°$, find x.

7. If $\sin(2x+20)° = \cos 40°$, find x.

8. If $\cos(2x-25)° = \sin 55°$, find x.

13.6 SAS Sine Formula for Area of a Triangle

Model Problem

Find the area of triangle PQR, shown below, in square units.

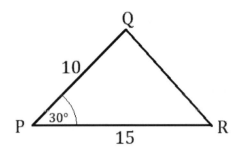

Solution:

(A) $Area = \frac{1}{2}ab \sin C$

(B) $= \frac{1}{2}(10)(15) \sin 30°$

(C) $= 37.5$

Explanation of steps:

(A) Write the formula $Area = \frac{1}{2}ab \sin C$

(B) Substitute the measures of the known angle *[30°]* for C and the adjacent sides *[10 and 15]* for a and b.

(C) Simplify, using the calculator to find the sine of the angle.

Practice Problems

1. In $\triangle ABC$, $a = 6$, $b = 8$, and $\sin C = \frac{1}{4}$. Find the area of $\triangle ABC$.

2. In $\triangle ABC$, $a = 12$, $b = 15$, and $m\angle C = 150°$. Find the area of $\triangle ABC$.

3. Find the area of △ABC to the nearest tenth of a square unit.

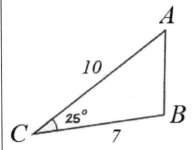

4. Find the area of △ABC, shown below.

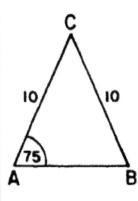

5. The triangle has two sides of 14 inches and 16 inches, and the angle between the sides is 30°. Find the area of the triangle, in square inches.

6. One angle of a triangular garden measures 70°, and the sides of the garden including this angle are 11 feet and 13 feet. Find, to the *nearest integer*, the number of square feet in the area of the garden.

7. If the vertex angle of an isosceles triangle measures 30° and each leg measures 4, what is the area of the triangle?

8. What is the area of a triangle with consecutive sides of 4 and 5 and an included angle of 59°, rounded to the *nearest tenth*?

9. Two sides of a triangle measure 16 feet and 21 feet, and the included angle measures 58°. What is the area, to the *nearest tenth of a square foot*, of the triangle?

10. In $\triangle DEF$, $m\angle D = 40$, $DE = 12$ meters, and $DF = 8$ meters. Find the area of $\triangle DEF$ to the nearest tenth of a square meter.

11. The diagram below shows the floor plan for a backyard, including a triangular deck represented by scalene triangle ABC. Find the area of the deck to the *nearest tenth of a square foot*.

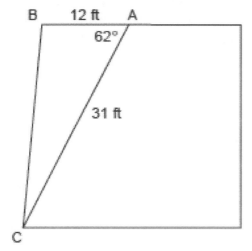

12. Find the area, to the *nearest tenth of a square foot*, of the isosceles triangle shown below.

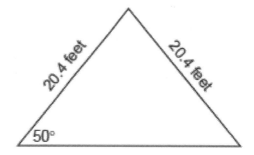

13. Find the area, to the *nearest integer*, of △ SRO shown below.

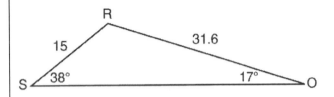

14. In △ ABC, m∠C = 30 and a = 8. If the area of the triangle is 12, what is the length of side b?

15. In △ ABC, m∠C = 30 and a = 24. If the area of the triangle is 42, what is the length of side b?

16. In △ ABC, m∠B = 30 and side a = 6. If the area of the triangle is 12, what is the length of side c?

Chapter 14. Quadrilaterals

14.1 Angles of Polygons

Model Problem

In the regular nonagon shown below, find x, the measure of exterior $\angle MBC$.

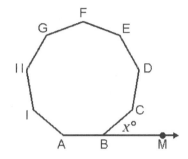

Solution:
$$x = \frac{360}{9} = 40°$$

Explanation:
Use the formula $x = \dfrac{360}{n}$, where n is the number of sides of the regular polygon.

Practice Problems

1. How many diagonals may a pentagon have?	2. How many diagonals may a decagon have?

193

3. What is the sum of the measures of the interior angles of an decagon?

4. Find the measure of each interior angle of a regular octagon.

6. Complete the table below for the measures of an exterior angle of regular polygons. Round to the *nearest tenth of a degree*.

Regular Polygons

Number of sides	Measure of an exterior angle
3	
4	
5	
6	
7	
8	
9	
10	

14.2 Properties of Quadrilaterals

MODEL PROBLEM

Find the measures of the numbered angles in the parallelogram.

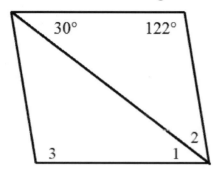

Solution:
(A) m∠1 = 30
(B) m∠2 = 28
(C) m∠3 = 122

Explanation of steps:
(A) In a parallelogram, opposite sides are parallel. Therefore, a diagonal acts as a transversal forming alternate interior angles that are congruent
 [∠1 and the 30° angle are alternate interior angles].
(B) In a parallelogram, consecutive angles are supplementary.
 [The whole angle consisting of ∠1 + ∠2 is supplementary to the 122° angle. Since m∠1 = 30, m∠2 = (180 − 122) − 30 = 28.]
(C) In a parallelogram, opposite angles are congruent.
 [∠3 is opposite the 122° angle.]

PRACTICE PROBLEMS

1. If the diagonals of a quadrilateral do not bisect each other, then the quadrilateral could be a (1) rectangle (3) square (2) rhombus (4) trapezoid	2. Which quadrilateral has diagonals that always bisect its angles and also bisect each other? (1) rhombus (3) parallelogram (2) rectangle (4) isosceles trapezoid
3. Which quadrilateral does *not* always have congruent diagonals? (1) isosceles trapezoid (3) rhombus (2) rectangle (4) square	4. Given three distinct quadrilaterals, a square, a rectangle, and a rhombus, which quadrilaterals must have perpendicular diagonals? (1) the rhombus, only (2) the rectangle and square (3) the rhombus and square (4) the rectangle, rhombus, and square

5. Which figure can serve as a counterexample to the statement, "If a quadrilateral has a pair of parallel sides and a pair of congruent sides, then the quadrilateral is a parallelogram."

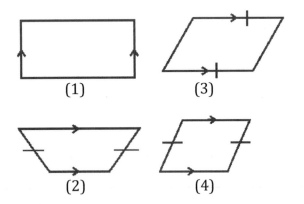

6. In parallelogram MNOP, the diagonals intersect at A and AO = 10. Find AM.

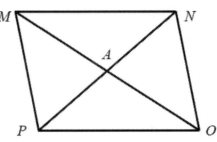

7. ABCD is a parallelogram. m∠A = 6x − 30 and m∠C = 4x + 10. Find m∠A.

8. In rhombus ABCD, AB = 8 and AC = 10. Find BD to the nearest tenth.

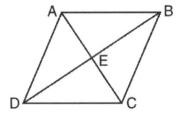

9. In the diagram below of rhombus ABCD, the diagonals $\overline{AC}$ and $\overline{BD}$ intersect at E. If AC = 18 and BD = 24, what is the length of one side of rhombus ABCD?

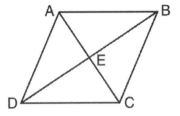

10. In rhombus ABCD, with diagonals $\overline{AC}$ and $\overline{DB}$, AD = 10. If the length of $\overline{AC}$ is 12, what is the length of $\overline{DB}$?

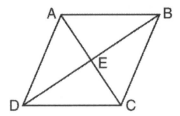

11. In the diagram below of rhombus *ABCD*, m∠*C* = 100. What is m∠*DBC*?

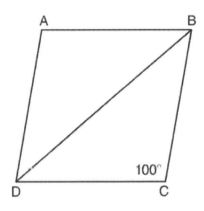

12. In the diagram below of parallelogram *ABCD* with diagonals $\overline{AC}$ and $\overline{BD}$, m∠1 = 45 and m∠*DCB* = 120. What is the measure of ∠2?

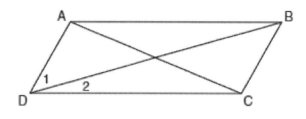

13. In the diagram below, *MATH* is a rhombus with diagonals $\overline{AH}$ and $\overline{MT}$. If m∠*HAM* = 12, what is m∠*AMT*?

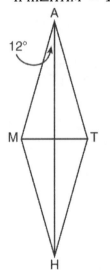

14. In the diagram of a rectangle below, the diagonals are drawn. If m∠1 = 42, what is m∠2?

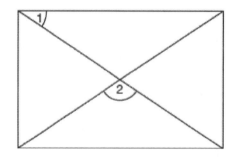

15. *ABCD* is a rhombus, m∠*BAC* = 30, and *AD* = 24. Find *DE*.

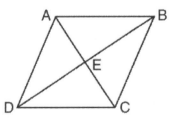

16. *ABCD* is a rhombus, m∠*ABC* = 60, and *AE* = 18. Find *DC*.

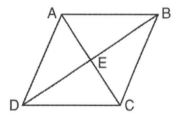

14.3 Trapezoids

Model Problem

The cross section of an attic is in the shape of an isosceles trapezoid, as shown in the accompanying figure. If the height of the attic is 9 feet, BC = 12 feet, and AD = 28 feet, find the length of $\overline{AB}$ to the *nearest foot*.

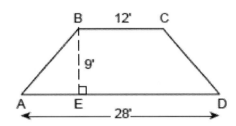

Solution:

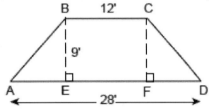

(A)

(B) $AE = \dfrac{28-12}{2} = 8$

(C) $(AB)^2 = 8^2 + 9^2 = 145$

$AB = \sqrt{145} \approx 12$

Explanation of steps:

(A) By drawing an altitude $\overline{CF}$, we know $\overline{CF} \cong \overline{BE}$. Since it is an isosceles triangle, we also know $\overline{AB} \cong \overline{CD}$. So, $\triangle ABE \cong \triangle DCF$ by HL. Therefore, $\overline{AE} \cong \overline{DF}$ by CPCTC.

(B) BCFE is a rectangle, so $EF = 12$. Since $AE = DF$, we find AE by subtracting $AD - EF$ and dividing by 2.

(C) The legs of $\triangle ABE$ are 8 and 9, so we can find the hypotenuse AB by using the Pythagorean Theorem.

Practice Problems

1. Given isosceles trapezoid ABCD with legs $\overline{AD}$ and $\overline{BC}$.

   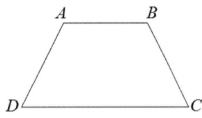

   (a) If $AC = 25$, find BD.

   (b) If m∠A = 105, find m∠D.

2. In the diagram below, LATE is an isosceles trapezoid with $\overline{LE} \cong \overline{AT}$, $LA = 24$, $ET = 40$, and $AT = 10$. Altitudes $\overline{LF}$ and $\overline{AG}$ are drawn. What is the length of $\overline{LF}$?

   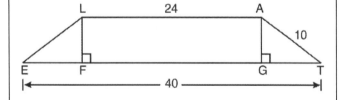

3. In the diagram below of isosceles trapezoid ABCD, AB = CD = 25, AD = 26, and BC = 12. What is the length of an altitude of the trapezoid?

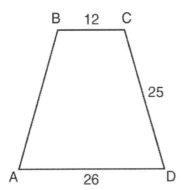

4. In isosceles trapezoid ABCD, $\overline{AB} \cong \overline{CD}$. If BC = 20, AD = 36, and AB = 17, what is the length of the altitude of the trapezoid?

5. The diagram shows a ramp, $\overline{RA}$, leading up to a level platform, $\overline{AM}$. The ramp forms an angle of 45° at its base. If platform $\overline{AM}$ measures 2 feet and is 6 feet above the ground, find the length of $\overline{RA}$.

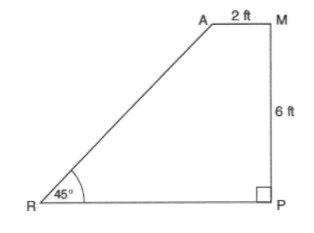

6. In trapezoid RSTV with bases $\overline{RS}$ and $\overline{VT}$, diagonals $\overline{RT}$ and $\overline{SV}$ intersect at Q. If trapezoid RSTV is not isosceles, name another triangle in the diagram which is equal in area to △ RSV.

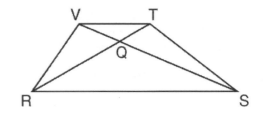

14.4 Use Quadrilateral Properties in Proofs

Model Problem

Given: Parallelogram *DEFG* with side $\overline{DE}$ extended to *H* such that ∠*DGK* ≅ ∠*EFH*.
Prove: $\overline{DK}$ ≅ $\overline{EH}$

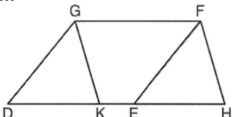

Solution:

Statements	Reasons
Parallelogram *DEFG*, ∠*DGK* ≅ ∠*EFH*	Given
$\overline{DG}$ ≅ $\overline{EF}$	opposite sides of a parallelogram are congruent
$\overline{DG}$ ∥ $\overline{EF}$	opposite sides of a parallelogram are parallel
∠*D* ≅ ∠*FEH*	corresponding angles theorem
△ *DGK* ≅ △ *EFH*	ASA
$\overline{DK}$ ≅ $\overline{EH}$	CPCTC

Practice Problems

1. Given: *ABCD* is a parallelogram
   Prove: △ *ABD* ≅ △ *CDB*

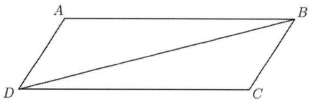

2. Given: Parallelogram NORB, diagonals $\overline{BO}$ and $\overline{NR}$ intersect at X
   Prove: △ BNX ≅ △ ORX

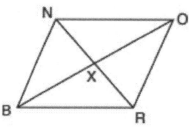

3. Given: Rectangle ABCD, M is the midpoint of $\overline{AD}$
   Prove: $\overline{BM} \cong \overline{CM}$

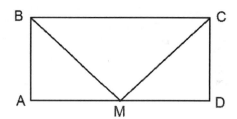

4. Given: Parallelogram FLSH, diagonal $\overline{FGAS}$, $\overline{LG} \perp \overline{FS}$, $\overline{HA} \perp \overline{FS}$
   Prove: △LGS ≅ △HAF

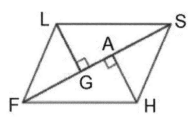

5. Given: Square ABCD with points E and F on $\overline{BC}$, $\overline{BE} \cong \overline{FC}$
   Prove: $\overline{AF} \cong \overline{DE}$

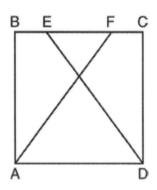

Quadrilaterals   14.4 Use Quadrilateral Properties in Proofs

6. Given: Rectangle $ABCD$ with points E and F on side $\overline{AB}$, $\overline{CE}$ and $\overline{DF}$ intersect at G, $\angle ADG \cong \angle BCG$
   Prove: $\overline{AE} \cong \overline{BF}$

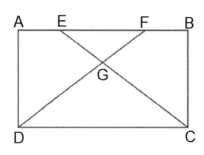

7. Given: Rhombus $EPRO$, $\overline{SEO}$, $\overline{PEV}$, $\angle SPR \cong \angle VOR$
   Prove: $\overline{SE} \cong \overline{EV}$

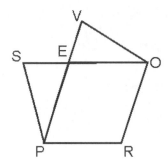

14.5 Prove Types of Quadrilaterals

Model Problem

Given: ∠UQV ≅ ∠RVQ and ∠TUQ ≅ ∠SRV
Prove: QRVU is a parallelogram

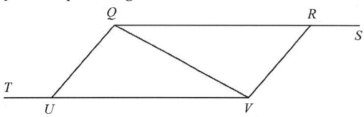

Solution:

Statements		Reasons
∠UQV ≅ ∠RVQ, ∠TUQ ≅ ∠SRV	(A)	Given
$\overline{UQ} \parallel \overline{VR}$		Alternate Interior Angles Converse
∠TUQ and ∠QUV are supplementary, ∠SRV and ∠VRQ are supplementary		Linear pairs are supplementary
∠QUV ≅ ∠VRQ	(A)	Supplements of congruent angles are congruent
$\overline{QV} \cong \overline{QV}$	(S)	Reflexive Property
△QUV ≅ △VRQ		AAS
$\overline{UQ} \cong \overline{VR}$		CPCTC
QRVU is a parallelogram		A quadrilateral with a pair of opposite sides both parallel and congruent is a parallelogram

Practice Problems

1. Based on the markings, determine if the figure is a parallelogram. If so, justify your answer.

   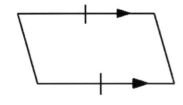

2. Based on the markings, determine if the figure is a parallelogram. If so, justify your answer.

   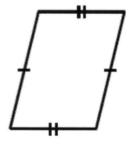

3. Which reason could be used to prove that a parallelogram is a rhombus?

    (1) Diagonals are congruent.

    (2) Opposite sides are parallel.

    (3) Diagonals are perpendicular.

    (4) Opposite angles are congruent.

4. Given: $\triangle AOB \cong \triangle COD$
   Prove: $ABCD$ is a parallelogram

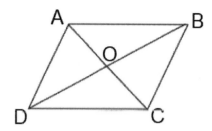

5. Given: $ABCD$ is a parallelogram, $DF = EB$
   Prove: $AECF$ is a parallelogram

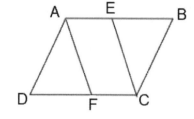

6. Given: ABCD is a parallelogram, and CEBF is a rhombus
   Prove: ABCD is a rectangle

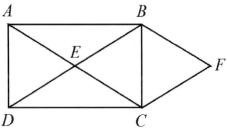

7. Given: Quadrilateral ABCD, diagonal $\overline{AFEC}$, $\overline{AE} \cong \overline{FC}$, $\overline{BF} \perp \overline{AC}$, $\overline{DE} \perp \overline{AC}$, $\angle 1 \cong \angle 2$
   Prove: ABCD is a parallelogram

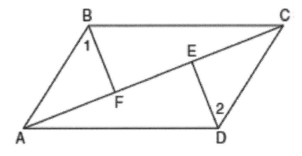

Chapter 15. Circles

15.1 Circumference and Rotation

Model Problem

To use the machine below, you turn the crank, which turns the pulley wheel, which winds the rope and lifts the box. Through how many rotations, to the *nearest tenth*, must you turn the crank to lift the box 10 feet?

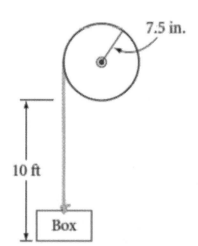

(not drawn to scale)

Solution:
$D = 10$ ft. $= 120$ in.
$C = 2\pi r = 15\pi$ in.
$R = \dfrac{D}{C} = \dfrac{120 \text{ in}}{15\pi \text{ in}} \approx 2.5$ rotations

Explanation:
Use the formula $R = \dfrac{D}{C}$

Practice Problems

1. A wheel has a radius of 5 feet. What is the minimum number of *complete* revolutions that the wheel must make to roll at least 1,000 feet?	2. A manufacturer wants to frame the circumferences of 8 inch diameter clocks with decorative rope. How many clocks can be framed from 100 feet of rope?

3. A surveyor uses a trundle wheel to measure the length of a trail. The device uses a 2-foot diameter wheel and counts the number of revolutions the wheel makes. If the device reads 1,100.5 revolutions at the end of the trail, how many miles long is the trail, to the *nearest tenth of a mile*? *(1 mile = 5,280 feet)*

4. The London Eye Ferris wheel has a diameter of 394 ft. If the Ferris wheel was to separate from its base and start rolling away, how many rotations, to the *nearest tenth*, would it take to travel the two miles to London Bridge? *(1 mile = 5,280 feet)*

5. On Pam's bicycle, each time she pushes the pedals to rotate 360°, the tires rotate three times. The diameter of a tire on the bicycle is 2 feet. What is the minimum number of *complete* rotations of the pedals needed for the bicycle to travel at least 1 mile? *(1 mile = 5,280 feet)*

15.2 Arcs and Chords

Model Problem

Given: $\overline{CB}$ is a diameter of circle A and $m\angle CAD = 70°$.

Find: $m\stackrel{\frown}{CD}$, $m\angle CBD$, $m\angle BAD$, $m\stackrel{\frown}{BD}$, and $m\angle ADB$.

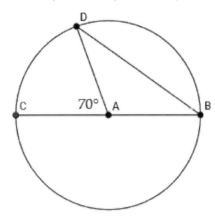

Solution:

(A) $m\stackrel{\frown}{CD} = m\angle CAD = 70°$

(B) $m\angle CBD = \frac{1}{2} m\stackrel{\frown}{CD} = 35°$

(C) $m\angle BAD = 180 - 70 = 110°$

(D) $m\stackrel{\frown}{BD} = m\angle BAD = 110°$

(E) $m\angle ADB = m\angle CBD = 35°$

Explanation of steps:

(A) A central angle is equal in measure to its intercepted arc.

(B) An inscribed angle is half the measure of its intercepted arc.

(C) A linear pair adds to 180°.

(D) A central angle is equal in measure to its intercepted arc.

(E) Base angles of an isosceles triangle are equal in measure. [$\overline{AB}$ and $\overline{AD}$ are radii.]

Practice Problems

1. In the diagram below, quadrilateral *JUMP* is inscribed in a circle.

   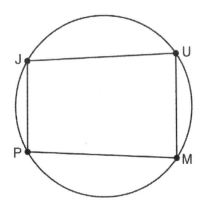

   Opposite angles *J* and *M* must be
   (1) right
   (2) complementary
   (3) congruent
   (4) supplementary

2. In circle *O*, diameter $\overline{AB}$ intersects chord $\overline{CD}$ at *E*. If *CE* = *ED*, then ∠*CEA* is which type of angle?

   (1) straight
   (2) obtuse
   (3) acute
   (4) right

3. Find the measure of the angle formed by the hands of the clock.

4. In circle *M* below, m∠*AMC* = 60°. Find the measure of inscribed ∠*ABC*.

   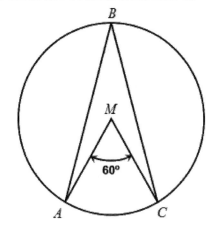

5. In the diagram below of circle O, m∠ABC = 24. What is the m∠AOC?

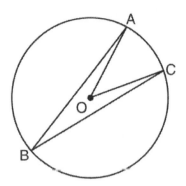

6. In the diagram below, $\overline{AB}$ is a diameter of circle O, and chord $\overline{AC}$ is drawn. If m∠BAC = 70°, then what is m $\widehat{AC}$?

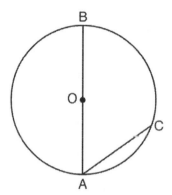

7. In the diagram of circle O below, chord $\overline{CD}$ is parallel to diameter $\overline{AOB}$ and m $\widehat{CD}$ = 110. What is m $\widehat{DB}$?

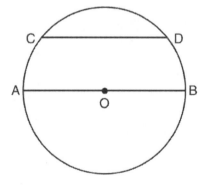

8. Trapezoid ABCD, with bases $\overline{AB}$ and $\overline{DC}$, is inscribed in circle O, with diameter $\overline{DC}$. If m $\widehat{AB}$ = 80, find m $\widehat{BC}$.

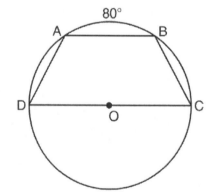

9. $\overline{BC}$ is a diameter of the circle below. If m∠B = 55°, find m∠C.

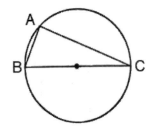

10. Quadrilateral DEFG is inscribed in a circle and m∠D = 86.
    (a) Find m$\widehat{GFE}$.
    (b) Find m∠F.

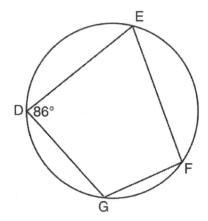

11. Quadrilateral ABCD is inscribed in a circle. If m$\widehat{AB}$ = 132° and m$\widehat{BC}$ = 82°, find m∠ADC.

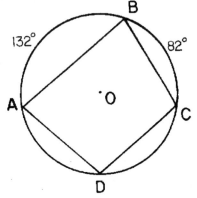

12. In the diagram of circle O, chords $\overline{DF}$, $\overline{DE}$, $\overline{FG}$, and $\overline{EG}$ are drawn such that m$\widehat{DF}$: m$\widehat{FE}$: m$\widehat{EG}$: m$\widehat{GD}$ = 5 : 2 : 1 : 7. Identify one pair of inscribed angles that are congruent to each other and give their measure.

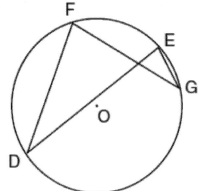

15.3 Tangents

A **tangent** is a straight line that touches a circle at only one point, called the **point of tangency**.

Model Problem

$\overline{XZ}$ is a diameter of circle O. Chords $\overline{XY}$ and $\overline{YZ}$ and tangent $\overleftrightarrow{WZ}$ are drawn.
m∠YXZ = 50°. Find m∠YZW.

Solution:
- (A) m∠XYZ = 90°
- (B) m∠XZY = 40°
- (C) m∠XZW = 90°
- (D) m∠YZW = 50°

Explanation of steps:
- (A) ∠XYZ is an inscribed angle of a semicircle
  [m∠XYZ = $\frac{1}{2}$ · 180° = 90°].
- (B) The sum of the measures of the angles of a triangle is 180° [180 − (90 + 50) = 40].
- (C) The diameter of a circle is perpendicular to the tangent passing through its endpoint.
- (D) Subtract [m∠XZW − m∠XZY = 90° − 40° = 50°].

Practice Problems

1. Circle O is inscribed in △ ABC with D, E, and F as points of tangency as shown. If AE = 15, EC = 10, and BF = 14, find the perimeter of △ ABC.

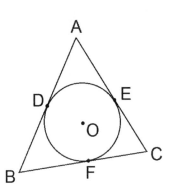

2. $\overline{CB}$ is tangent to circle A at B. Find the length of $\overline{CD}$.

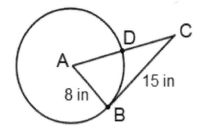

3. From external point A, two tangents to circle O are drawn. The points of tangency are B and C. Chord $\overline{BC}$ is drawn to form △ABC. If m∠ABC = 66, what is m∠A?

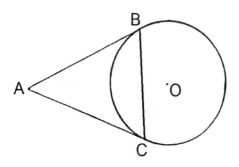

4. $\overline{AC}$ and $\overline{BC}$ are tangent to circle O at A and B, respectively, from external point C. If m∠ACB = 38, what is m∠AOB?

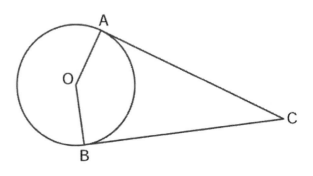

5. Rays QR and QS are tangents to circle P at R and S. If m∠Q = 54°, find m $\widehat{RS}$.

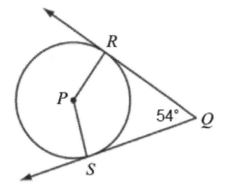

6. $\overleftrightarrow{ED}$ and $\overleftrightarrow{EC}$ are tangents to circle B at D and C. Find m∠DEC.

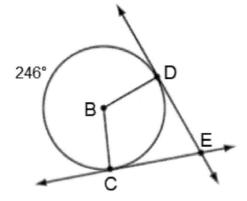

7. To the two circles below, sketch all common tangent lines.

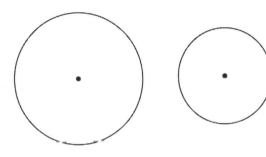

8. Circles A and B are tangent at point C and $\overline{AB}$ is drawn. Sketch all common tangent lines.

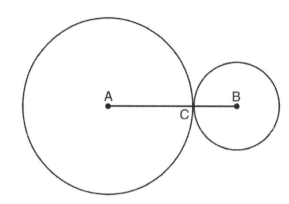

15.4 Secants

Model Problem

In the accompanying diagram, $\overline{PAB}$ and $\overline{PCD}$ are secants drawn to circle O. If $PA = 8$, $PB = 20$, and $PD = 16$, find PC.

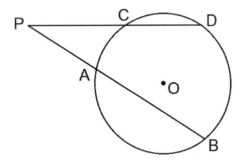

Solution:
$(PA)(PB) = (PC)(PD)$
$(8)(20) = (PC)(16)$
$PC = 10$

Explanation of steps:
Use the formula $a(a + b) = c(c + d)$ where a and c are the external parts of the secants [PA and PC] and b and d are the internal parts of the secants [chords AB and CD].

Practice Problems

1. What is the name of each line or segment in relation to circle O?

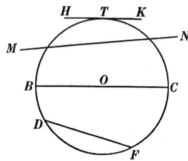

a) $\overleftrightarrow{HK}$

b) $\overleftrightarrow{MN}$

c) $\overline{BC}$

d) $\overline{DF}$

2. Find x.

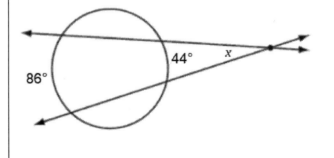

3. In circle O, $\overline{PAC}$ and $\overline{PBD}$ are secants. If m $\widehat{CD} = 70$ and m $\widehat{AB} = 20$, what is the measure of $\angle P$?

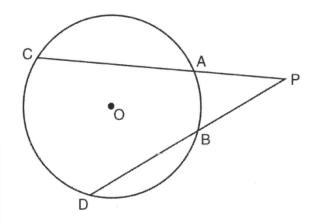

4. Find m $\widehat{DE}$.

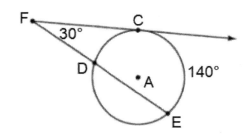

5. $\overline{PA}$ is tangent to circle O at A, secant $\overline{PBC}$ is drawn, $PB = 4$, and $BC = 12$. Find PA.

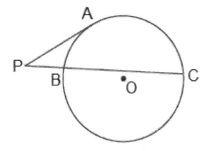

6. Secants $\overline{PQR}$ and $\overline{PST}$ are drawn to a circle from point P. If $PR = 24$, $PQ = 6$, and $PS = 8$, determine and state the length of $\overline{PT}$.

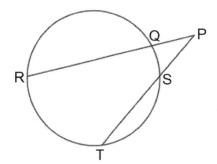

7. $\overline{CBA}$ and $\overline{CED}$ are secants. $AC = 12$, $BC = 3$, and $DC = 9$. Find EC.

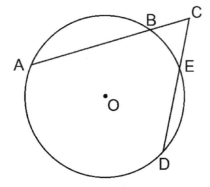

8. $\overline{PS}$ is a tangent to circle O at point S and $\overline{PQR}$ is a secant. If $PS = x$, $PQ = 3$, and $PR = x + 18$, find x.

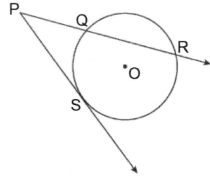

(Not drawn to scale)

9. In circle O, chords $\overline{RT}$ and $\overline{QS}$ intersect at M. Secant $\overline{PTR}$ and tangent $\overline{PS}$ are drawn to circle O. RM is two more than TM, $QM = 2$, $SM = 12$, and $PT = 8$.

a) Find the length of $\overline{RT}$.

b) Find the length of $\overline{PS}$.

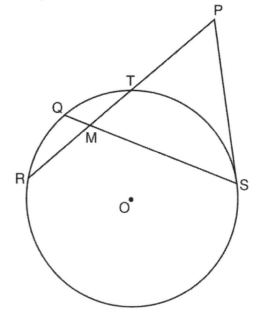

10. In circle O, chords $\overline{AB}$ and $\overline{CD}$ intersect at E. Secant $\overline{FDA}$ and tangent $\overline{FB}$ are drawn to circle O from external point F and chord $\overline{AC}$ is drawn. $m\overset{\frown}{DA} = 56$, $m\overset{\frown}{DB} = 112$, and the ratio of $m\overset{\frown}{AC} : m\overset{\frown}{CB} = 3 : 1$.

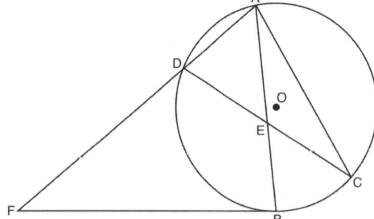

a) Find m∠CEB.

b) Find m∠F.

c) Find m∠DAC.

11. In the diagram of circle B below, radius $\overline{BC}$, tangent $\overline{AC}$, and secant $\overline{APBQ}$ are drawn, forming right triangle ABC. Let a and b represent the lengths of the legs BC and AC, respectively, and let c represent the length of the hypotenuse, AB. Use the theorems taught in this chapter to show that $a^2 + b^2 = c^2$.
Hint: $\overline{BP}$ and $\overline{BQ}$ are also radii.

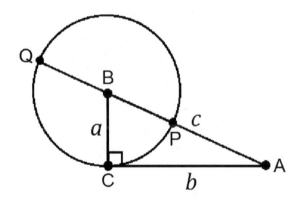

219

15.5 Circle Proofs

Model Problem

Given: Circle O, $\overline{DB}$ is tangent to the circle at B, $\overline{BC}$ and $\overline{BA}$ are chords, and C is the midpoint of $\widehat{ACB}$.

Prove: $\angle ABC \cong \angle CBD$

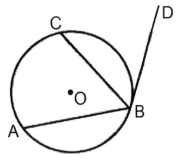

Solution:

Statements	Reasons
Circle O, $\overline{DB}$ is tangent to the circle at B, $\overline{BC}$ and $\overline{BA}$ are chords, and C is the midpoint of $\widehat{ACB}$	Given
$m\widehat{AC} = m\widehat{BC}$	Definition of midpoint
$m\angle ABC = \frac{1}{2} m\widehat{AC}$	The measure of an inscribed angle is one-half the measure of its intercepted arc
$m\angle CBD = \frac{1}{2} m\widehat{BC}$	The measure of an angle formed by a tangent and a chord is one-half the measure of the intercepted arc
$\frac{1}{2} m\widehat{AC} = \frac{1}{2} m\widehat{BC}$	Multiplication property of equality
$m\angle ABC = m\angle CBD$	Substitution
$\angle ABC \cong \angle CBD$	Definition of congruence

Circles   15.5 Circle Proofs

PRACTICE PROBLEMS

1. Given: Circle O, $\overarc{AB} \cong \overarc{AC}$
   Prove: $\triangle AOC \cong \triangle AOB$

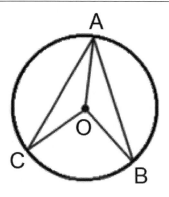

2. Given: Circle Q, $\overline{PQR} \perp \overline{ST}$
   Prove: $\overline{PS} \cong \overline{PT}$

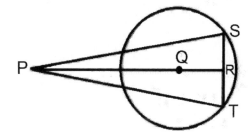

221

3. Given: Tangents $\overline{PA}$ and $\overline{PB}$, radii $\overline{OA}$ and $\overline{OB}$, and $\overline{OP}$ intersects the circle at C.
   Prove: $\angle AOP \cong \angle BOP$.

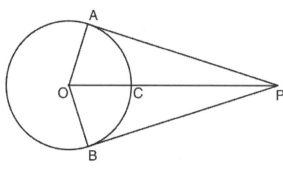

4. Given: Diameter $\overline{BOD}$, m $\widehat{BR} = 70$, and m $\widehat{YD} = 70$.
   Prove: $\triangle RBD \cong \triangle YBD$

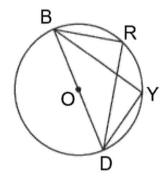

5. Given: Quadrilateral ABCD is inscribed in circle O, $\overline{AB} \parallel \overline{DC}$, and diagonals $\overline{AC}$ and $\overline{BD}$ are drawn.
   Prove: $\triangle ACD \cong \triangle BDC$.

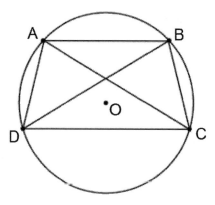

6. Given: In circle O, $\overline{AD}$ is a diameter with $\overline{AD}$ parallel to chord $\overline{BC}$, chords $\overline{AB}$ and $\overline{CD}$ are drawn, and chords $\overline{BD}$ and $\overline{AC}$ intersect at E.
   Prove: $\overline{BE} \cong \overline{CE}$

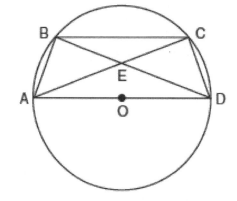

15.6 Arc Lengths and Sectors

Model Problem

In the diagram below, find the length of arc *AB*, to the *nearest tenth*.

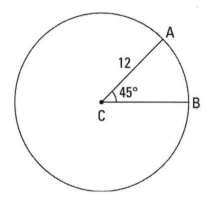

Solution:

(A) $\dfrac{\theta}{360°} = \dfrac{L}{2\pi r}$

(B) $\dfrac{45°}{360°} = \dfrac{L}{2\pi \cdot 12}$

(C) $360L = 1080\pi$

(D) $L = 3\pi \approx 9.4$

Explanation of steps:
(A) Write the proportion for the length of an arc.
(B) Substitute known values.
(C) Cross multiply.
(D) Solve.

Practice Problems

1. Find the length of $\widehat{AB}$.

   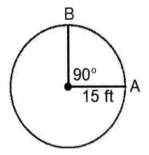

2. Find the length of $\widehat{AB}$.

   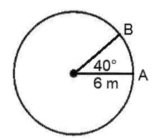

3. Diameter *AC* is 34 cm long. Find the length of $\widehat{BC}$ to the *nearest centimeter*.

   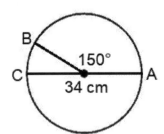

4. Find the measure of the central angle of an arc if its length is 14π and the radius is 18.

5. The length of $\widehat{AS}$ is 247 feet. Find, to the *nearest degree*, the measure of $\widehat{AS}$.

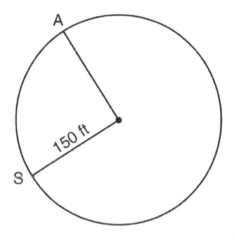

6. An object travels along a circular path from point A to point B as shown in the diagram below. The radius of the circle is 2.40 meters and the central angle intercepting $\widehat{AB}$ is 165°. How far does the object travel along the path, to the *nearest tenth of a meter*?

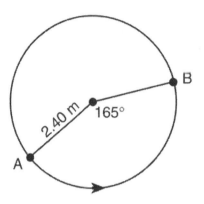

7. Find the area of the shaded sector.

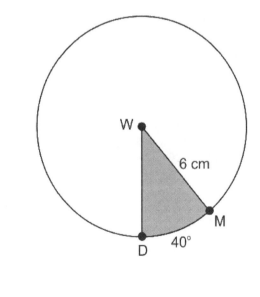

8. Find the area of the shaded sector.

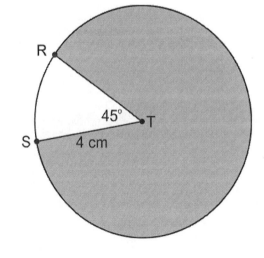

9. Find the area of the shaded segment.

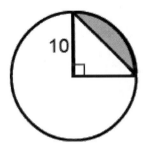

10. Find the area of the shaded segment.

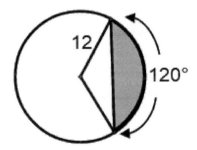

11. Two cities, *H* and *K*, are located on the same line of longitude and the difference in the latitude of these cities is 9°, as shown in the diagram below. If Earth's radius is 3,954 miles, how many miles north of city *K* is city *H* along arc *HK*? Round your answer to the *nearest tenth of a mile*.

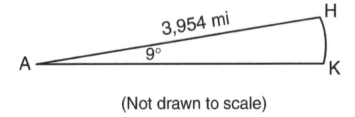

(Not drawn to scale)

12. In the diagram below, circle L has a radius of 6 and circle S has a radius of 3. The two circles intersect at points A and B, where $\overline{AB}$ is a diameter of circle S.

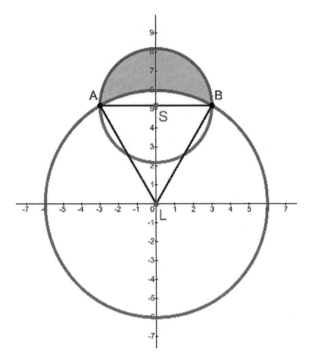

a) Find $m\angle ALB$.

b) Find the exact area of the segment of circle L formed by chord $\overline{AB}$.

c) Find the exact area of the shaded region of circle S.

13. In the diagram below, circle L has a radius of $3\sqrt{2}$ and $SB = SL = 3$. The two circles intersect at points A and B, where $\overline{AB}$ is a diameter of circle S.

Show that the area of the shaded region of circle S is equal to the area of △ ABL.

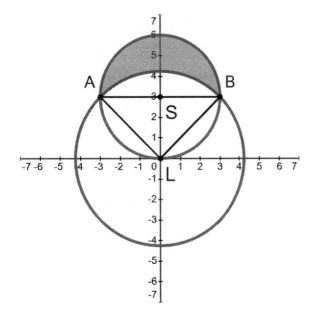

Chapter 16. Solids

16.1 Volume of a Sphere

Model Problem

A baseball has a diameter of $\frac{9}{\pi}$ inches. Find the volume of the baseball, rounded to the *nearest hundredth of a cubic inch*.

Solution:
(A) $r = \frac{9}{2\pi}$
(B) $V = \frac{4}{3}\pi r^3 = \frac{4}{3}\pi \cdot \left(\frac{9}{2\pi}\right)^3$ cubic inches
(C) $V \approx 12.31$ cubic inches.

Explanation of steps:
(A) The radius is half the diameter.
  $[r = \frac{1}{2} \cdot \frac{9}{\pi} = \frac{9}{2\pi}]$
(B) Use the formula $V = \frac{4}{3}\pi r^3$.
  [Substitute $\frac{9}{2\pi}$ for r.]
(C) Find the approximate volume using the calculator.

Practice Problems

1. A sphere has a diameter of 18 meters. Find the *exact* volume of the sphere, in cubic meters.	2. A basketball has a diameter of $\frac{29.5}{\pi}$ inches. Find the volume of the basketball, to the *nearest tenth*.

3. A balloon in the shape of a sphere has a diameter of 6 inches. An air pump inflates the balloon further until it has a diameter of 12 inches. By how many cubic inches, *in terms of π*, did the balloon increase in volume?

16.2 Volume of a Prism or Cylinder

Model Problem

A water tank in the shape of a rectangular prism measures 6 feet by 5 feet by 4 feet and is completely filled with water. Drew needs to fill as many barrels as he can with water from the tank. Each barrel is shaped as a cylinder with a 1 foot radius and a height of 2 feet. What is the maximum number of whole barrels can he completely fill with water taken from the tank?

Solution:
(A) $V_{tank} = lwh = (6)(5)(4) = 120$ cubic feet
$V_{barrel} = \pi r^2 h = \pi(1^2)(2) = 2\pi \approx 6.28$ cubic feet
(B) Since $120 \div 6.28 \approx 19.11$, Drew can fill 19 whole barrels.

Explanation of steps:
(A) Calculate the volume of each solid by substituting for the variables in the appropriate formulas *[the tank is a rectangular prism and the barrel is a cylinder]*.
(B) State the solution *[divide the volumes to determine how many barrels can be filled]*.

Practice Problems

1. What is the volume, in cubic feet, of the rectangular prism shown below?

   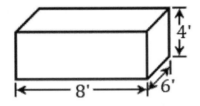

2. Each edge of a cube measures 1.5 cm. What is the volume of the cube in cubic centimeters?

3. What is the volume of this cylinder, to the nearest hundredth of a cubic inch?

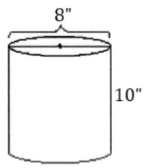

4. What is the volume of this cylinder, to the nearest tenth of a cubic inch?

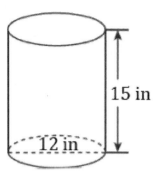

5. The volume of a cylindrical can is 32π cubic inches. If the height of the can is 2 inches, what is its radius, in inches?

6. A cardboard box has length $x - 2$, width $x + 1$, and height $2x$. Write the volume of the box as a polynomial in terms of x.

7. How many cubes with 5-inch sides will completely fill a cube that is 10 inches on a side?

8. A cylindrical soup can has a volume of 342 cm³ and a diameter of 6 cm. Express the height of the can in terms of π.

   Determine the maximum number of soup cans that can be stacked on their base between two shelves if the distance between the shelves is exactly 36 cm.

9. The diagram below shows the dimensions of two water tanks. The larger tank is completely filled with water. Then, water is taken from the larger tank to fill the smaller tank. How many cubic inches of water remain in the larger tank?

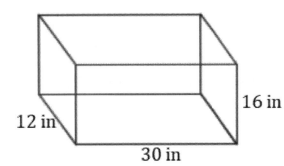 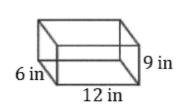

10. In the accompanying diagram, a rectangular container with the dimensions 10 inches by 15 inches by 20 inches is to be filled with water, using a cylindrical cup whose radius is 2 inches and whose height is 5 inches. What is the maximum number of full cups of water that can be placed into the container without the water overflowing the container?

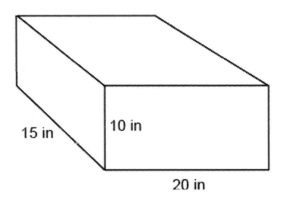
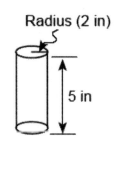

11. Freezer Fresh offers ice cream in two different containers. One container is shaped as a rectangular prism with dimensions of 5 inches by 3.5 inches by 7 inches. The other container is shaped as a cylinder with a diameter of 5 inches and a height of 7 inches.

Which container holds more ice cream? How much more ice cream, to the *nearest tenth of a cubic inch*, does the larger container hold?

Solids 16.3 Volume of a Pyramid or Cone

16.3 Volume of a Pyramid or Cone

MODEL PROBLEM

A regular pyramid with a square base is shown below. A side, s, of the base of the pyramid is 12 meters, and the height, h, is 42 meters. What is the volume of the pyramid in cubic meters?

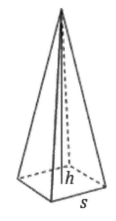

Solution:
$$V = \frac{1}{3}(12)^2(42) = 2{,}016 \text{ m}^3$$

Explanation of steps:
The volume of a pyramid is $V = \frac{1}{3}Bh$. Find the area of the base, B, and then substitute for B and h in the formula. Calculate the result.
[Since the base is square, the area of the base, $B = s^2 = 12^2$.]

PRACTICE PROBLEMS

1. What is the volume of this rectangular pyramid, in cubic inches?

   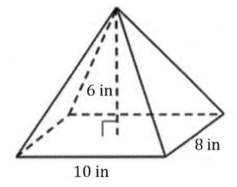

2. The volume of a cone is 8 cm³. What is the volume of a cylinder with the same base as the cone and the same height as the cone?

3. A paper container in the shape of a right circular cone has a radius of 3 inches and a height of 8 inches. Determine and state the number of cubic inches in the volume of the cone, in terms of π.

4. A regular pyramid has a height of 12 centimeters and a square base. If the volume of the pyramid is 256 cubic centimeters, how many centimeters are in the length of one side of its base?

5. The solid below is made up of a cube and a square pyramid. Find its volume.

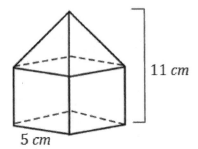

6. A container consisting of a cylinder on top of a cone is shown below.

a) Find the volume of the container, to the *nearest thousandth*.

b) If the container is filled to half of its capacity (i.e., half of its volume) with water, how high, from the apex of the cone, will the water level reach, to the *nearest tenth*?

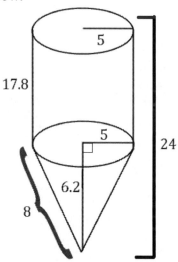

16.4 Density

Model Problem

The Longball baseball factory needs to limit the content weight of a case of baseballs to a maximum of 10 pounds. A baseball is a sphere with a diameter of 2.9 inches and a density of 0.4 ounces per cubic inch. Determine and state the maximum number of baseballs that a case may contain.

Solution:
(A) $V = \frac{4}{3}\pi r^3 = \frac{4}{3}\pi(1.45)^3 \approx 12.77 \text{ in}^3$
(B) $W = VD = 12.77 \text{ in}^3 \cdot \frac{0.4 \text{ oz}}{1 \text{ in}^3} \cdot \frac{1 \text{ lb}}{16 \text{ oz}} \approx 0.32 \text{ lb}$
(C) $\frac{10}{0.32} = 31.25$, so a maximum of 31 baseballs may be packed in a case.

Explanation of steps:
(A) Find the volume of the object. *[For one baseball, use the formula for a sphere, substituting the radius 1.45, which is half the diameter.]*
(B) Determine the weight of the object as the product of the volume and the density. *[We also need to include a conversion fraction here, to convert from ounces to pounds, since the maximum load for a case of baseballs is given in terms of pounds.]*
(C) State the solution. *[Divide 10 lbs. by the weight of each baseball.]*

Practice Problems

1. A rectangular prism, measuring 5 cm by 10 cm by 2 cm, is made of lead, which has a density of 11.34 grams per cubic cm. Find the weight of the item, in grams.

   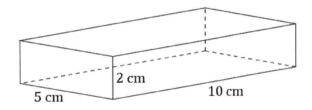

2. A 10 cm³ sample of copper weighs 88.6 g. What is the density of copper?

Solids 16.4 Density

3. A sample of iron in the shape of a rectangular prism has dimensions of 2 cm by 3 cm by 2 cm. If the sample weighs 94.44 g, what is the density of iron?

4. Gasoline has a density of 0.7 g/mL. Find the volume, in mL, of 9.8 g of gasoline.

5. A triangular pyramid, with a right triangle as its base, is shown below. The pyramid is made of a plastic which has a density of 2 g/cm³. Find the weight, in pounds, of the pyramid, to the *nearest tenth of a pound*.
   1 inch = 2.54 cm
   1 kg = 2.2 lbs

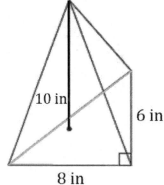

16.5 Lateral Area and Surface Area

Model Problem

How many square centimeters of vinyl would it take to completely cover the surface of the cylinder shown below, to the *nearest square centimeter*?

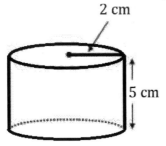

Solution:
(A) $SA = 2\pi r^2 + 2\pi rh$
(B) $= 2\pi(2)^2 + 2\pi(2)(5)$
(C) $= 8\pi + 20\pi = 28\pi$
 ≈ 88 sq cm

Explanation of steps:
(A) Write the appropriate formula for the surface area of the given solid *[cylinder]*.
(B) Substitute the given values for the variables *[r = 2 and h = 5]*.
(C) Simplify, and round the result.

Practice Problems

1. What is the surface area, in square feet, of the rectangular prism shown below? 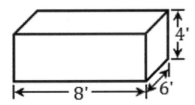	2. What is the surface area of a rectangular prism measuring 3 feet long, 1.5 feet wide, and 2 feet high?

3. Billy wants to wrap a gift box that has a length of 3.0 cm, a width of 2.2 cm, and a height of 7.5 cm. How many square centimeters of wrapping paper will he need to entirely cover the box?

4. A rectangular prism has a length of 5.5 cm, a width of 3 cm, and a height of 6.75 cm. Find the surface area of the prism, in square centimeters.

5. Find the volume and surface area of the rectangular prism below.

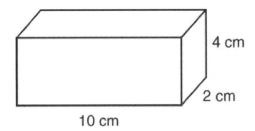

6. The volume of a cube is 64 cubic inches. What is its total surface area, in square inches?

7. What are the lateral area and surface area, in terms of π, of the cylinder shown below?

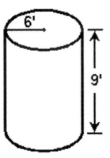

8. A cylinder has a height of 11 feet and a radius of 5 feet. What is the surface area, in square feet, of the cylinder, to the *nearest tenth*?

9. A rectangular prism has a length of $x + 3$, a width of $x - 4$, and a height of 5. Represent the surface area of the prism as a trinomial in terms of x.

10. Three different cylinders, with radii of 2, 4, and 6, have equal volumes of 144π each. Make a table showing the radius, height and surface area of the three cylinders. Which cylinder has the least surface area?

Solids

16.6 Rotations of Two-Dimensional Objects

Model Problem

Describe the object formed when the right triangle to the right is rotated about the y-axis. What is the volume of the object?

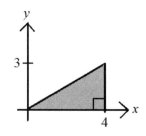

Solution:
(A) The object is a cylinder with a cone cutout, as shown below.

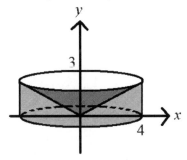

(B) $V_{\text{object}} = V_{\text{cylinder}} - V_{\text{cone}} = \pi r^2 h - \frac{1}{3}\pi r^2 h = \frac{2}{3}\pi r^2 h = \frac{2}{3}\pi (4)^2 (3) = 32\pi$

Explanation of steps:
(A) Visualize the rotation of the figure.
[You can imagine a rectangle (below) rotating around the axis, forming a cylinder, and the darker triangle rotating around the axis, forming a cone cutout.]

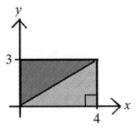

(B) To find the volume of a solid with a cutout, subtract the volume of the cutout from the solid.
[For both the cylinder and cone, the radius is 4 and the height is 3.]

Practice Problems

1. What type of object is formed when the circle is rotated about the line? 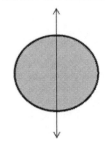	2. What type of object is formed when the metal flag, shown as a right triangle below, spins around its pole? 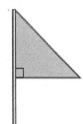

3. What type of object is formed when the square is rotated about the line?

   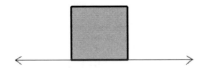

   If each side of the square measures 5 inches, find the volume of the object formed by the rotation.

4. Describe the object formed when the figure below is rotated about the line.

   What is the volume of the object?

   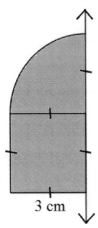

   3 cm

5. The area between two concentric semicircles is shaded below. Describe the object formed when the shaded figure is rotated about their extended diameter. What is the volume of the object?

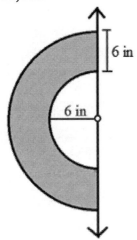

6. The arrow figure below consists of an isosceles triangle with legs 13 units long and a 5 by 12 unit rectangle. The height of the figure is 24 units.

Describe the object formed by rotating the figure around its axis of symmetry. What is the volume of the object?

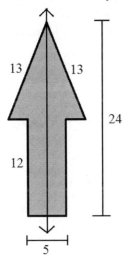

16.7 Cross Sections

Model Problem

A cube with sides of length $\sqrt{8}$ is cut by a plane that passes through three of the cube's vertices as shown. What type of triangle is the cross section? What is the area of the triangle?

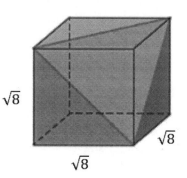

Solution:

(A) $c^2 = (\sqrt{8})^2 + (\sqrt{8})^2 = 16$
$c = 4$
Cross section is an equilateral triangle with sides of length 4.

(B) Draw a height, h, of the triangle, which bisects a base.

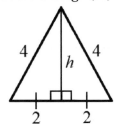

$h^2 + 2^2 = 4^2$
$h = \sqrt{12} = 2\sqrt{3}$
$Area = \frac{1}{2}bh = \frac{1}{2}(4)(2\sqrt{3}) = 4\sqrt{3}$

Explanation of steps:

(A) The sides are hypotenuses of congruent isosceles right triangles *[with legs of $\sqrt{8}$]*, so the sides are congruent and the triangle is equilateral. We can find the length of a side using the Pythagorean Theorem.

(B) The height of an equilateral triangle bisects the triangle into two congruent right triangles. We can find the height using the Pythagorean Theorem, and then calculate the area.
[An alternative method of finding the area of the triangle is to use the SAS sine formula,
$Area = \frac{1}{2}ab \sin C = \frac{1}{2}(4)(4) \sin 60° = 4\sqrt{3}.$ *]*

Practice Problems

1. The cross section of the cone shown below is a (1) triangle (3) semicircle (2) square (4) circle 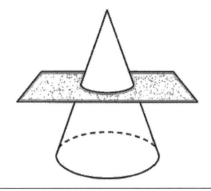	2. The cross section of the square pyramid shown below is a (1) triangle (3) rectangle (2) square (4) trapezoid 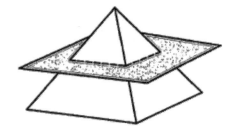

3. The cross section of the triangular pyramid shown below is a

   (1) triangle (3) rectangle
   (2) square   (4) trapezoid

4. The cross section of the square pyramid shown below is a

   (1) triangle (3) rectangle
   (2) square   (4) trapezoid

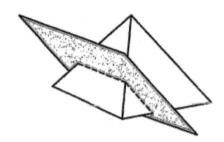

5. The cross section of the rectangular prism shown below is a

   (1) triangle (3) rectangle
   (2) square   (4) hexagon

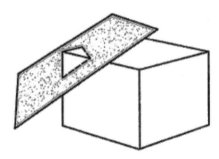

6. What is the shape of the cross section when a right pentagonal prism, shown here, is cut by a plane that is

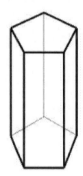

   a) parallel to the pentagonal bases?

   b) perpendicular to the bases?

Chapter 17. Constructions

17.1 Copy Segments, Angles, and Triangles

MODEL PROBLEM

Construct a triangle with sides of lengths *a*, *b*, and *c*, as shown below. Be sure the longest side of your triangle lies on $\overline{PQ}$ and that point *P* is one of the triangle's vertices.

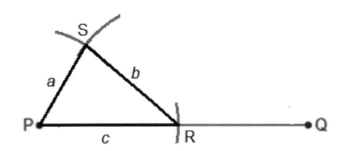

Solution:

Explanation of steps:
- (A) Copy segment *c* onto $\overline{PQ}$ with *P* as an endpoint. Label the other endpoint *R*.
- (B) Set the compass to the length of *a* and draw an arc from *P*.
- (C) Set the compass to the length of *b* and draw an arc from *R*. Label the intersection of the arcs point *S*.
- (D) Use the straightedge to draw $\overline{PS}$ and $\overline{RS}$.

Constructions   17.1 Copy Segments, Angles, and Triangles

PRACTICE PROBLEMS

1. Construct $\overline{EF}$ such that $EF = AB + CD$.

   A•————•B   C•————————•D

2. Construct a copy of right angle *ABC*.

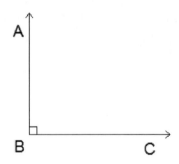

247

3. Construct a triangle that is congruent to equilateral triangle ABC.

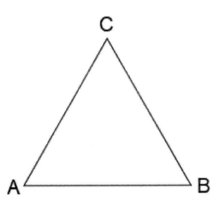

4. Based on the construction marks, which congruence rule was used to construct this triangle as congruent to a given triangle, not shown?

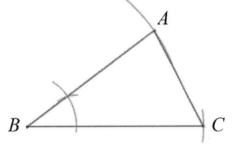

5. Based on the construction marks, which congruence rule was used to construct this triangle as congruent to a given triangle, not shown?

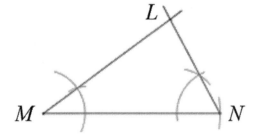

17.2 Construct an Equilateral Triangle

MODEL PROBLEM

Construct an equilateral triangle with $\overline{RS}$ as one side.

Solution:

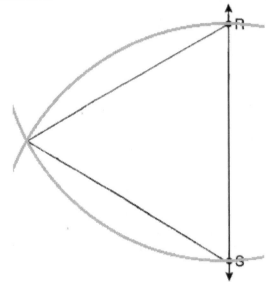

Explanation of steps:
- (A) Set the compass to the length of $\overline{RS}$.
- (B) Draw an arc from R and an arc from S with radii equal to RS.
- (C) Using the intersection of the arcs as the third vertex, draw the sides.

Constructions   17.2 Construct an Equilateral Triangle

PRACTICE PROBLEMS

1. Construct equilateral triangle *ABC* given $\overline{AB}$ below.

    A•————————————•B

2. Construct a 60° angle with its vertex at point *P*.

    •P

3. Construct a 120° angle with its vertex at point *P*. *Hint: Construct two adjacent 60° angles.*

    •P

250

17.3 Construct an Angle Bisector

MODEL PROBLEM

Construct the bisector of the angle shown below.

Solution:

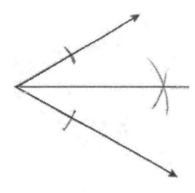

Explanation of steps:
- (A) From the vertex and using the same radius, draw arcs onto both rays.
- (B) From the intersection of one arc with a ray, draw an arc midway between the rays.
- (C) From the intersection of the other arc and ray, and using the same radius, draw another arc midway between the rays.
- (D) Use a straightedge to draw a line from the vertex through the point of intersection of the two arcs.

PRACTICE PROBLEMS

1. Construct the bisector of this angle.

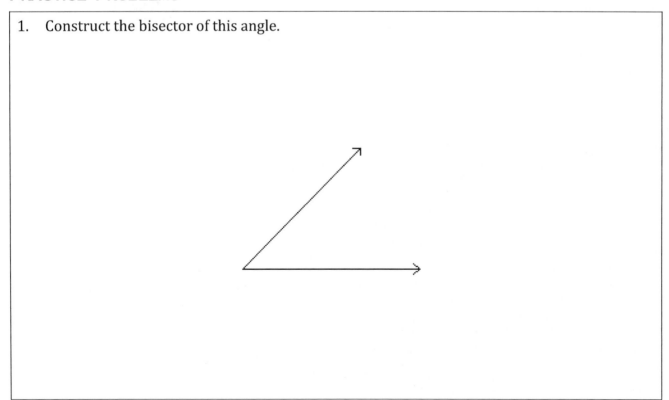

2. Construct the bisector of this angle.

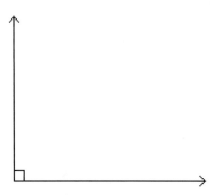

3. Construct the bisector of ∠ABC.

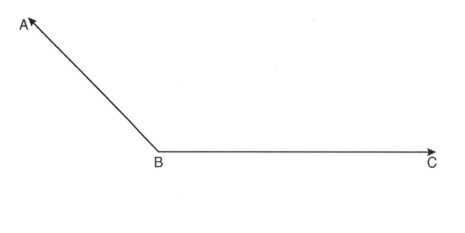

4. Construct the bisector of ∠XYZ.

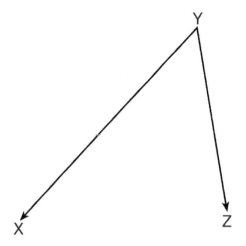

5. Construct the bisector of ∠ABC.

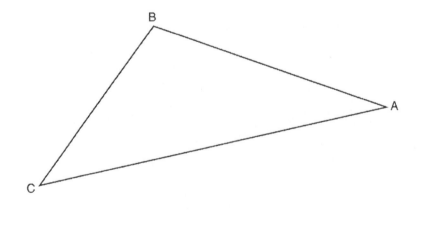

Constructions   17.3 Construct an Angle Bisector

6. Construct the bisector of ∠CDE.

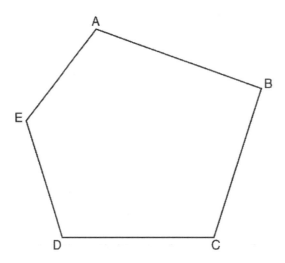

7. Construct an equilateral triangle with $\overline{AB}$ as a side. Using this triangle, construct a 30° angle with its vertex at A.

A •————————————• B

Constructions   17.4 Construct a Perpendicular Bisector

17.4 Construct a Perpendicular Bisector

MODEL PROBLEM

Construct a 45° angle.

Solution:

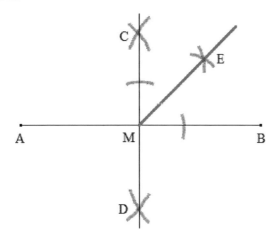

Explanation of steps:

(A) Draw $\overline{AB}$. Construct $\overline{CD}$, the perpendicular bisector of $\overline{AB}$. Label point M, the midpoint of $\overline{AB}$. m∠CMB = 90°.

(B) Construct $\overline{ME}$, the bisector of ∠CMB. m∠EMB = $\frac{1}{2}$m∠CMB = 45°.

PRACTICE PROBLEMS

1. Construct the perpendicular bisector of $\overline{AB}$.

2. Construct the perpendicular bisector of $\overline{AB}$. Label the midpoint of $\overline{AB}$ as point M.

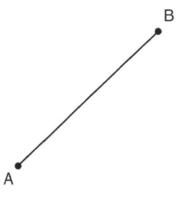

3. On the diagram of $\triangle ABC$, construct the perpendicular bisector of side $\overline{AC}$.

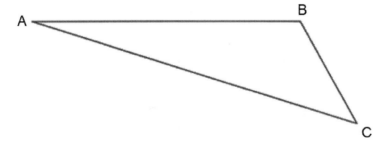

4. On the diagram of △ PQR, construct the median $\overline{RS}$.

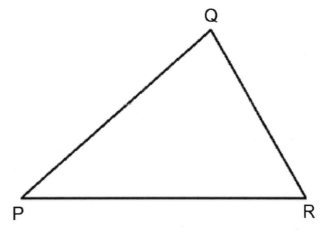

5. Use construction methods to divide $\overline{XY}$ into four congruent parts.

17.5 Construct Lines Through a Point

Model Problem

Given segments of lengths a and b below, construct a trapezoid WXYZ in which $WZ = a$, $XY = b$, and $\overline{WZ} \parallel \overline{XY}$.

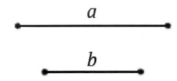

Solution:

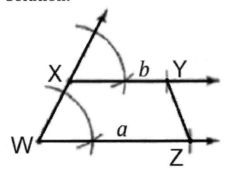

Explanation of steps:

(A) Construct $\overline{WZ}$ with length a.
(B) Draw point X not on $\overline{WZ}$. Then draw $\overrightarrow{WX}$.
(C) Construct a ray through X parallel to $\overrightarrow{WZ}$ by copying congruent corresponding angles.
(D) Construct $\overline{XY}$ with length b, then draw $\overline{YZ}$.

Practice Problems

1. The diagram below shows the construction of $\overleftrightarrow{AB}$ through point P parallel to $\overleftrightarrow{CD}$.

   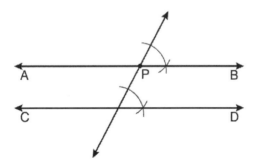

   Which theorem justifies this method of construction?

   (1) If two lines in a plane are perpendicular to a transversal at different points, then the lines are parallel.

   (2) If two lines in a plane are intersected by a transversal to form congruent corresponding angles, then the lines are parallel.

   (3) If two lines in a plane are intersected by a transversal to form congruent alternate interior angles, then the lines are parallel.

   (4) If two lines in a plane are intersected by a transversal to form congruent alternate exterior angles, then the lines are parallel.

2. Construct the line through point P that is perpendicular to $\overline{AB}$.

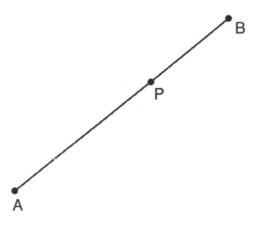

3. Construct the line through point P that is perpendicular to $\overline{AB}$.

4. Construct the altitude of △ ABC from point C to side $\overline{AB}$.

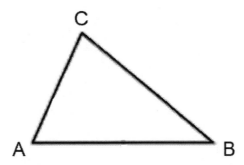

5. Construct the line through point P that is parallel to the given line.

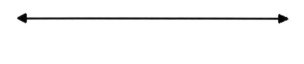

P

6. Match each construction with one of the diagrams below.
    ____ a) angle bisector
    ____ b) perpendicular through a point on a line
    ____ c) perpendicular from a point to a line
    ____ d) perpendicular bisector
    ____ e) equilateral triangle
    ____ f) altitude of a triangle

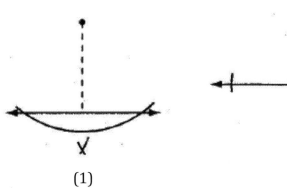

(1)

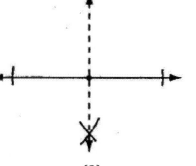

(2)

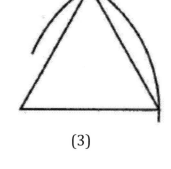

(3)

(4)

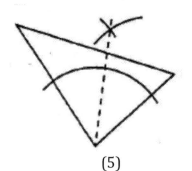

(5)

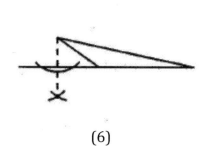

(6)

17.6 Construct Inscribed Regular Polygons

Model Problem

A dodecagon is a twelve-sided polygon. Construct a regular dodecagon inscribed within a circle.

Solution:

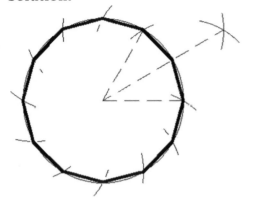

Explanation of steps:

A dodecagon has twice as many sides as a hexagon.
- (A) Set the compass to the radius of the circle and draw six equidistant marks of this length around the circle.
- (B) Draw the radii to two consecutive marks and bisect the angle between them, which will also bisect the arc between them.
- (C) Reset the compass to the radius of the circle and place the tip at the point of intersection of the angle bisector and the circle.
- (D) Draw six additional equidistant marks around the circle.
- (E) The twelve marks represent the vertices of the dodecagon. Draw the chords between them.

Practice Problems

1. Construct an equilateral triangle inscribed within the circle below.

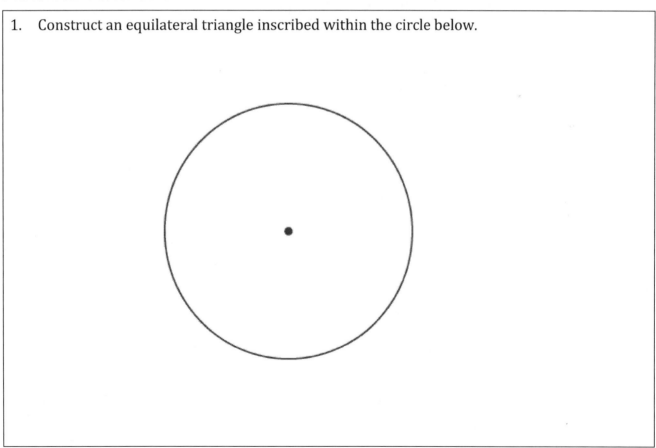

2. Construct a square inscribed within the circle below.

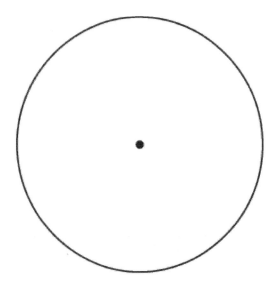

3. Inscribe a regular octagon inside the circle below.
   Hint: an octagon has twice as many sides as a square.

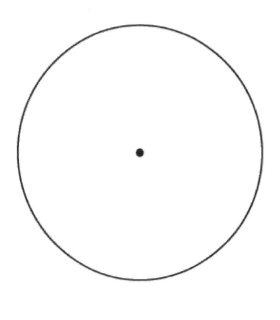

17.7 Construct Points of Concurrency

Model Problem

Use construction to determine whether the triangle below is acute, right or obtuse.

Solution:

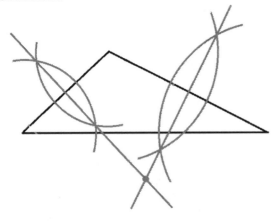

The circumcenter lies outside the triangle, so the triangle is obtuse.

Explanation of steps:
The circumcenter lies inside an acute triangle but outside an obtuse triangle. For a right triangle, the circumcenter is the midpoint of the hypotenuse. Therefore, we can determine the type of triangle by locating the circumcenter, which is the intersection of the perpendicular bisectors of the sides.

Instead, we could have located the orthocenter, which is the intersection of the altitudes of the triangle. In the same way, the orthocenter lies inside an acute triangle but outside an obtuse triangle. For a right triangle, the orthocenter is the vertex of the right angle.

Practice Problems

1. Construct the circumcenter of triangle *ABC* and label it *P*.

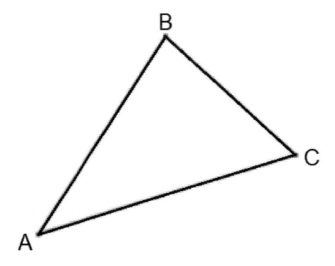

2. Construct the orthocenter of triangle *DEF* and label it *P*.

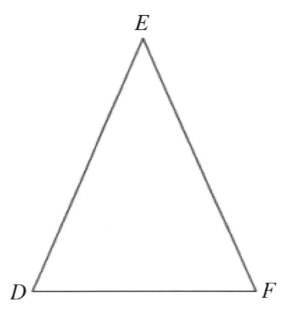

17.8 Construct Circles of Triangles

MODEL PROBLEM

Point C is both the incenter and circumcenter of the triangle below.
a) Inscribe a circle inside the triangle.
b) Circumscribe a circle around the triangle.

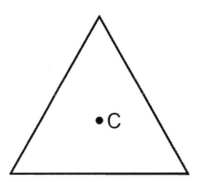

Solution:

(A)

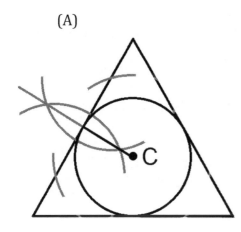

(B)

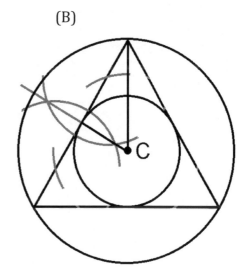

Explanation of steps:
(A) To inscribe a circle, first find the radius by constructing a line perpendicular to a side through the incenter. The radius is the distance from the incenter to the side. Using this radius and the incenter as the center, draw the circle with a compass.
(B) To circumscribe a circle, first find the radius, which is the distance from the circumcenter to one of the vertices. *[A radius is drawn to the top vertex.]* Using this radius and the circumcenter as the center, draw the circle with a compass.

Practice Problems

1. Which principle is demonstrated by the construction below?

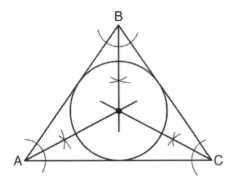

   (1) The intersection of the angle bisectors of a triangle is the center of the inscribed circle.

   (2) The intersection of the angle bisectors of a triangle is the center of the circumscribed circle.

   (3) The intersection of the perpendicular bisectors of the sides of a triangle is the center of the inscribed circle.

   (4) The intersection of the perpendicular bisectors of the sides of a triangle is the center of the circumscribed circle.

2. The diagram below shows a construction of the circle circumscribed about △ABC.

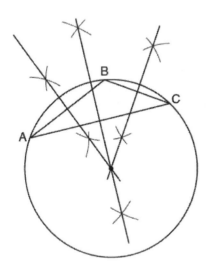

   This diagram shows that the center of the circle is found by constructing the intersection of

   (1) the angle bisectors of △ABC

   (2) the medians to the sides of △ABC

   (3) the altitudes to the sides of △ABC

   (4) the perpendicular bisectors of the sides of △ABC

3. Match each construction with one of the diagrams below.

_____ a) orthocenter

_____ b) incenter

_____ c) centroid

_____ d) circumcenter

_____ e) inscribed circle

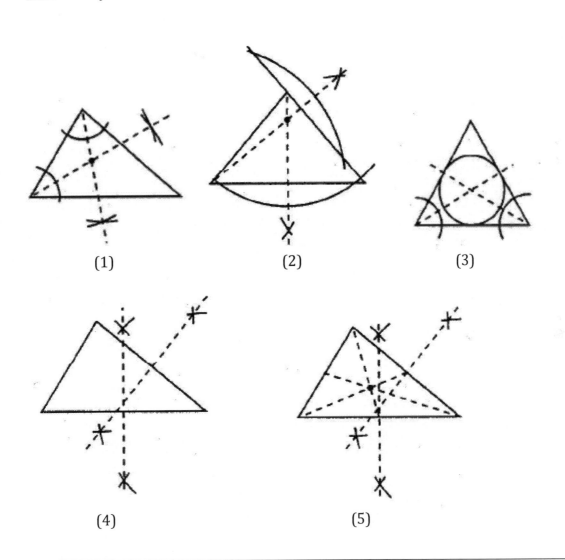

Made in the USA
Middletown, DE
21 August 2024

59533767R00150